Central Himalaya

Ecology, Environmental Resources and Developments

– *Author* –
D.D. Maithani

2022

Daya Publishing House®

A Division of

Astral International Pvt. Ltd.

New Delhi – 110 002

First Impression, 1991
Reprinted, 2022

ISBN: 9788170350996 (HB)

Publisher's Note:

Every possible effort has been made to ensure that the information contained in this book is accurate at the time of going to press, and the publisher and author cannot accept responsibility for any errors or omissions, however caused. No responsibility for loss or damage occasioned to any person acting, or refraining from action, as a result of the material in this publication can be accepted by the editor, the publisher or the author. The Publisher is not associated with any product or vendor mentioned in the book. The contents of this work are intended to further general scientific research, understanding and discussion only. Readers should consult with a specialist where appropriate.

Every effort has been made to trace the owners of copyright material used in this book, if any. The author and the publisher will be grateful for any omission brought to their notice for acknowledgment in the future editions of the book.

Published by : **Daya Publishing House®**
A Division of
Astral International Pvt. Ltd.
– ISO 9001:2015 Certified Company –
4736/23, Ansari Road, Darya Ganj,
New Delhi-110 002
Ph. 011-43549197, 23278134
E-mail: info@astralint.com
Website: www.astralint.com

PREFACE

The central Himalaya is an area of perennial interest both to the scholar and the layman. The scenic beauty and the grandeur of its physical form have invited interest since the very early stages of the people. Numerous references found in the very early text such as the Puranas and the literary works. Its pride like a crown, is the U.P. Himalaya— the abode of Gods, 'Devbhumi, or 'Tapobhumi' is how it finds mention in mythology and religious literature.

The region is well enriched with varied environmental resources. It is a land of difficult environment, significant resources and innumerable problems about which it is easy to generalise but not so easy to understand. Man has paid heavy ecological costs for blundering with nature's scheme. Attributes of civilisation have often been found bearing inverse relationship with world ecosystems. Indiscriminate exploitation of resources has caused serious ecological damage to the hilly tracts and the plains as well. Extensive suination of forests has reduced tree and other vegetative cover of the central Himalayan watershed areas by as much as 45% in the past 45 years.

Over-grazing, primitive practice of Jhumming agriculture, unplanned tourist resorts, road building, indiscriminate hunting and cultivation of sloping submarginal land, the various factors have all taken heavy toll of the natural environment. Landslides and other forms of damage are likely to increase progressively unless ecological concepts are brought to bear on the management of central Himalayan resources.

The region being very vast and varied and the problems so intricate, it is therefore, necessary to create an environmental consciousness among the masses. The development activities should be studied and their impact evaluated. Wherever feasible, steps should be taken to mitigate the adverse impact. This collection of papers, an attempt has been made, not only to integrate the studies into a unified whole but to accurately identify the relative importance of the subsequent papers. Part I depicts the ecological characteristics presenting details about the geomorphologic and Geo-ecological aspects. The Part II, comprised of three papers, depicts the environmental resources presenting details of recreational resources, mineral resources and development. Part III consists of four papers, the case studies, depicts the Rural settlements, Rural landuse pattern, agriculture and mountain development etc. This volume contains some thought provoking papers in order to arouse some environmental and ecological awareness and to promote interest in the Himalayan environment and ecosystem.

I am deeply indebted to Prof. S. P. Nautiyal, Vice-chancellor H.N.B. Garhwal University, Srinagar (Garhwal), Prof. K.P. Nautiyal, Deptt. of History,

Prof. N.S. Bisht, Deptt. of Economics and Dr. R.D. Gaur, Deptt. of Botany, for their paternal encouragement from time to time. In bringing out this book, my sincere thanks are due to Mr. Kedar Singh, Librarian, Delhi Administration, Delhi, all contributors and many others who promoted me to bring out the present volume.

Department of Geography
HNB Garhwal University
Srinagar, U.P. 246174
February 12, 1990.

D.D. Maithani

CONTENTS

LIST OF CONTRIBUTORS

1. Prof. S.P. Nautiyal,
Vice-Chancellor,
H.N.B. Garhwal University,
Srinagar (Garhwal).

2. Dr. R.D. Gaur,
Reader in Botany,
H.N.B. Garhwal University,
Srinagar (Garhwal).

3. Dr. A.K. Sandilya,
Reader in Geology,
H.N.B. Garhwal University,
Srinagar (Garhwal).

4. Dr. D.D. Maithani,
Reader in Geography
H.N.B. Garhwal University,
Srinagar (Garhwal).

5. Dr. U.S. Rawat,
Lecturer in Geology,
Government (P.G.) College,
Rishikesh (U.P.).

6. Dr. G.S. Rawat,
Reader in Geology,
H.N.B. Garhwal University,
Srinagar (Garhwal).

7. Mr. N.P. Naithani,
Department of Geology,
H.N.B. Garhwal University,
Srinagar (Garhwal).

8. Dr. G.S. Rajwar,
Lecturer in Botany,
Government (P.G.) College,
Rishikesh.

9. Dr. M.S. Anantharaman,
Department of Geology,
D.B.S. College,
Dehra Dun.

10. Dr. K.N. Joshi,
Reader in Economics,
H.N.B. Garhwal University,
Srinagar (Garhwal).

11. Dr. M.S. Negi,
Department of Geography,
H.N.B. Garhwal University,
Srinagar (Garhwal).

12. Mr. B.P. Naithani,
Lecturer in Geography,
H.N.B. Garhwal University,
Srinagar (Garhwal).

13. Mr. R.S. Negi,
Department of Geology,
H.N.B. Garhwal University,
Srinagar (Garhwal).

14. Dr. R.K. Verma,
Lecturer in Geography,
Government (P.G.) College,
Rishikesh.

Part I
Ecology

GROUND WATER PROVINCES OF U.P. HIMALAYA

S.P. Nautiyal and Arun K. Sandilya

The U.P. Himalaya is rich in scenic beauty, hazy and lush green mountain slopes, glaciers, river valleys, besides its own mineral wealth. The Himalaya being the perennial source of the water for most of the rivers including the Yamuna and the Ganga, which originate from the glaciers, the Yamunotri and Gangotri glaciers respectively. The Garhwal Himalaya comprises of mountaineous area, Himalayan ranges covering nearly 19,000 sq. kms. The mountain rises upto 7,500 m. height. The Ganga Yamuna Ramganga river system discharges about 10^3 m/sec quantity of the water and silt etc to the ocean. The erosion during the past 50 years, due to the indiscriminate forest felling seems to have been considerably accentuated causing serious depletion of the ground water, springs and increasing the volume of the flood water and silt in the drainage system. The Himalayas are unique mountainous area where human habitation sprawl upto a height of 2500 m. on the mountain slopes and upto then height about 3,500 m. along the valley i.e. in the Niti and Mana etc. The main occupations are agriculture, horticulture and animal husbandry.

Water is a vital need for human and animals existance for both drinking irrigation and agriculture. The valley have fair abundance of water in the main streams, which is utilized by the gravity flow or lifted by pumping to meet out the growing demands. On the mountain slopes springs area the main source, which are not only dispersed but the ecological desperadation has resulted in the depleting their discharges. Some of them go dry during the dry seasons, causing extreme hardship to the inhabitants. The density of the springs are one in each 10 km^2, thus the village people have to spend good deal of their time in fetching water like those of desert dwellers. In some area water has to be fetched from the valley usually 500 to 1000 m. below. The Government has launch on very expensive water supply schemes, which in many cases become infeasible due to high cost of maintenance in such rough terrain and despite the government efforts for the drinking water supply remains a problem.

So far no systematic studies have been carried out due to the lack of appreciation of the hardship of local populace and neglect, which has been an inheritance from British Raj. The inhabitant complains that the water discharge has gone down and many cases, once perennial springs become dry in summers. The possible causes could be the receding vegetation, pressure on the land, increase in consumption of water because of rise in the population, choking of the spring channels, diversion of the springs flow due to earth movements etc.

After independence, Government of India has planned few schemes for the hill areas in Garhwal, on the priority basis for the supply of drinking water, including the hydrum, lifting of the water (supply of water for drinking to the higher ridges) and some for the lift irrigation. The Jal Nigam and Jal Sansthan are engaged in supply of water for drinking and irrigation purposes.

Drainage System

The Garhwal Himalaya serves as the perennial reservoir of water for most of the rivers, which are of immense value to north India. The region is sell drained by numerous rivers and rivulets (locally known as Gad and Gadhera/Raula). The region has three main drainage systems, 1. The Ganga Drainage system 2. The Yamuna Drainage system. 3. Ramganga system . The whole of the Garhwal except the western part of Utterkashi and Dehradun Districts are drained by the Ganga system. The Bhagirathi and Alaknanda rivers which originated from the Gangotri and Badrinath/ Satopanth glaciers from the Chaukhamba peak (7135 m). After flowing in the opposite directions forming very deep gorges and narrow deep valleys with meandered channels makes the confluence at Davprayag, forming a garland. The Janhvi, the Bhagirathi, the Alaknanda, provides a excellent example of the antecedent drainage. The main tributaries of the Alaknanda are Mandakini (originate from Kedarnath Glaciers), Pinder (from Pindari glacier), Nandakini (Nandagiri glacier) and Dhauliganga (from the Nandadevi glacier) joins the stream Alaknanda at Rudraprayag, Karnprayag, Nandprayag and Vishnuprayag respectively. The Alaknanda and the Bhagirathi river are VII order stream.

The Yamuna has largest tributary Tons which drained the western part of the Garhwal Himalaya, boardering with Himachal in the west. The Yamuna originated from the Yamunotri glacier, lying on the south western slopes of the Bander-Punchh peak (6315 m.). The Tons originate from the northern slope of the Bander-Punchh and meet the Yamuna near Kalsi.

The drainage has been analysed with the help of the topographic sheets and landsat imageries of the area. It is noticed that each drainage basin of individual river system has typical drainage pattern, which is directly controlled by the lithology and structural configurations. The characteristic drainage pattern are dendritic, trellis, angular, rectangular, parallel, obsequent, consequent, centripital, radial, barbed, and braided (only is Srinagar and Tehri).

Glaciers

Most of the landforms of the Garhwal Himalaya show evidences of ice sculpture/glacial topography, resulting from the past gigantic glaciers. The valleys more than about 2000 m. height depicts glacier features, wherever they have not been completely obliterated by the fluvial actions. The glacier landforms are well preserved above an elevation of 3000 m. The important glaciers are given in (Table-1).

Table 1

Name of the Glaciers	Location	Length	Height in m.	River
1. Gangotri	30°45'-30°55'N : 79°5'-79°.15' E	30 km.	6614	Bhagirathi
2. Yamunotri	30°50'-30°00'N : —	10 km		Yamuna
3. Chourabari/ Kedarnath	30°50'-31°00'N : 79°00'-79°15' E	14 km.	6940	Mandakini
4. Badrinath/ Satopanth	30°41'-30°50'N : 79°10'-79°25'E	10 km		Alaknanda
5. Bahagat Kharak	30°48'-30°50'N : 79°15'-79°25'E	5.5 km.		
6. Khatling	- - -	-	-	Bhilangana
7. Nandadevi	- - -	-	7817	Dhauliganga
8. Arwa	- - -	-	-	-
9. Nandagiri	- - -	-	-	Nandakini
10. Pindari	- - -	-	-	Pinder
11. Kurantoli	- - -	-	-	-

Climate

The sufficient data are lacking over much of the area and in such an area of complicated mountainous terrain, microclimates are of considerable importance. There are not only variation but also complexities of climates as well as complexities of weather, accentuated by the relief of the land. There are variation of exposure to the sun light and rain bearing clouds have the effects of producing very intricate patterns of local climate. In the summer months, the valleys experience hot streamy tropical climate, while at a distance of about 75 km. the great Himalayan ranges bears the height of snow field of the world. In the winter months the valley have the heavy fog and the hill tops receives the snow fall. The precipitation of every localities are directly related not only to the altitudinal zones in which they exists, but also to its situations in relation to the ridge or overlapping spurs.

The monsoon commences towards the end of June and ceases by the mid of September. The winter depression causes snowfall for seven to eight days in each of the three months from January to March. April and May are rather marked by thunder and occasional hailstorms. During the months of May and first half of the June, before the break or the Monsoon conventional rains occurs in the afternoon in small amount (10-30 mm.). The south facing slopes are having high rate of precipitation than that of north facing slopes. The zone of maximum precipitation during both summer and winter, lies between 1000 to 2500 m. The area above the elevation of 2500 m. experience the small amount of the summer rains also.

TABLE 2: **Distribution of Rainfall and Snowfall in Garhwal**

S.No.	Station	Altitude in m.	Rainfall in mm.	Snowfall in mm.	Temp. Min.	Temp. Max.
1.	Srinagar	550	1000-1500	Nil	4-6°	40°
2.	Karnprayag	884	1500-2000	Nil	4-6°	25°
3.	Tehri	778	1500-2000	Nil	—	—
4.	Dehradun	682	1800-3500	Nil	—	—
5.	Narendranagar	1080	3000-4500	—	—	—
6.	Kotdwar	396	1800-3500	Nil	—	—
7.	Gopeshwar	—	2000-4500	10-100	0°	20°
8.	Utterkashi	—	2000-3500	10-100	0°	20°
9.	Mussourrie	1800	3100-5000	10-100	0°	20°
10.	Joshimath	2000	3000-5000	10-150	0°	18°
11.	Pauri	1700	2500-4500	—	—	—

Sources of Water

The Himalayan terrain is drained by numerous streams, rainfed and glacial-fed. In some of the Himalayan tracts, there are very few perennial streams and most of the catchment territory of such streams are dry, there is extreme scarcity of water even drinking water. The drinking water supply is generally made from lifting of water, which is quite costly affairs. The main sources of water in the Garhwal Himalaya are as follows—

1. Surface Water
   - (a) The Ganga River System
   - (b) The Yamuna River System and
   - (c) The Ram Ganga River System
2. Ground Water
   - (a) Springs
   - (b) Lakes and Tals
   - (c) Hot Water Springs
3. Glaciers and High altitudinal lakes

Ground Water Provinces of Himalaya

The Himalayan terrain consists of following principal formations of rock types, which are described here as ground water province (GWP) —

1. Alluvial Province
2. Siwalik Province
3. Schistose Province
4. Quartzite Province
5. Granite Crystalline and
6. Glacial Province

These above mentioned G.W. Provinces have their diagnostic characteristics.

1. Alluvial Province (Dun-Bhabar Alluvium Province)

In this province we have the Dun-valley fills and the Bhabar alluvial zone. This province comprises of the alluvial formation in the sub outer Himalaya, spread over area of about 1000 km^2, which is considered to be a good aquifer for good quantity of water can be utilized for the irrigation and domestic purposes.

The Dun Valley fills are characterised by the presence of large boulder, pebbles and cobbles, sand and silt etc. they are prolific and thick enough in the Dun Valley. The water yielding capacity of gravels are large and is easily support normal size of wells of capacity $2.5m^3/sec$ ($0. 6m^2/sec$). In the Dun valley the ground water table (G.W.T.) towards the foot of Lesser Himalayan ranges may be at the depth of about 100 m below the surface. But towards and after the Siwalik range the Ground water table is at shallow depth, at about 6 to 8 m depth. The water potential is great and the formations are amenable to only percussion drilling on account of boulder nature of the formations, similarly are the Bhabar Formation, where ground water table is at peiziometric head permits the over flowing wells, such as in Nainital, Haldwani and Bhabar areas. This province is shown with yellow colour in Pt.-1. The water can be collected through old river channels, for the supply of drinking water.

2. Siwalik Province

The Siwalik formations are characterised by the presence of sandstone, intercalate with shales, claystones etc. They show highly dissected topography and the water table is below the Ground Surface. The yields of such formations are fair, porosity of sandstone is high, but generally low because admixture of silt in sandstone. The claystone has extremely poor yield.

The Siwalik are also, prove to be a good aquifer, which contain the water upto the depth of 200-300 meter. Some amount of water is going waste in the form of springs/stream water, runoff etc. The total area of these formation is approx. about 1500 km^2 in the part of foot hills, between Yamuna on the West and Kali on the East. This province is demarcated by Main Boundary Thrust in the North.

3. Schistose Province

The areas covered by schistose rock mica schists, etc. in the part of Lesser Himalaya, contains water only along the foliation planes and are poor aquifer. It is in the such areas where there is a great scarcity of drinking water.

4. Quartzite Province: (Fractured Quartzite)

In the various fractured quartzite provinces the ground water conditions are slightly better than schistose province, because these rocks are highly fractured and jointed in which sufficient quantity of water along the secondary porosity are accumulated. The stored water comes out as a springs even at higher altitudes. Some amounts of water is stored along the folded quartzites of Garhwal group rocks, at their fold noses. It can not be a perennial source of water. The spring of these areas becomes dry in summers.

5. Limestone—Dolomite Province (Calcareous province)

The limestone - dolomite are shown in blue in the map. Through they do not have primary porosity but due to leaching by ground water have invariably small vugs to large caves. This short of conditions exists throughout the Himalaya and in Kashmir, the source of Jhelum is located in the Triassic limestone, where large volume of ground water emerges at Sheshnag.

The area covering calcareous formations (limestone + dolomite) are approx. 1500 Km^2 in the Lesser Higher Himalaya, part of U.P. Himalaya. These rocks have a good source of water, having a good aquifer with fairly large storage capacity and yields.

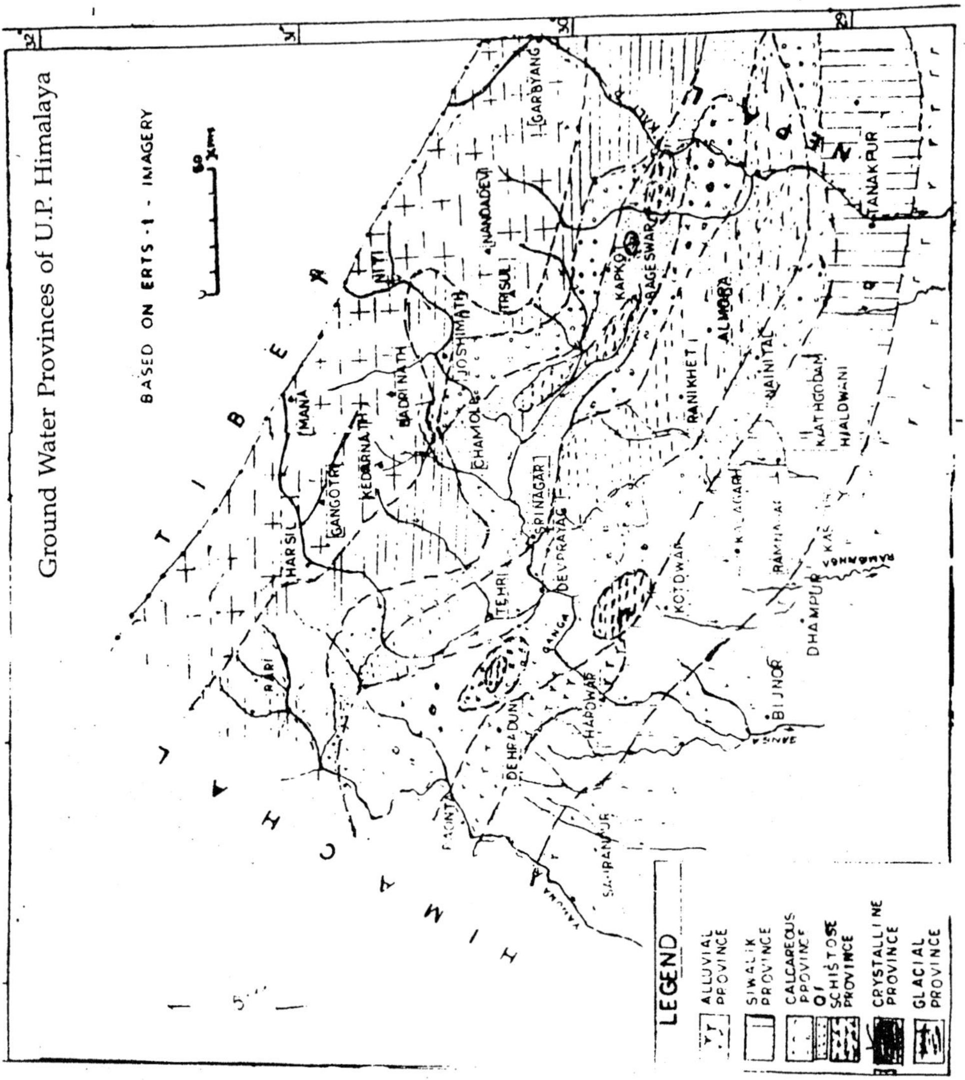
Ground Water Provinces of U.P. Himalaya
BASED ON ERTS-1 - IMAGERY
TIBET
GARBYANG
NANDADEVI
NITI
TRISUL
JOSHIMATH
BADRINATH
MANA
KEDARNATH
GANGOTRI
CHAMOLI
HARSIL
SRINAGAR
DEVPRAYAG
TEHRI
KAPKOT
BAGESWAR
ALMORA
RANIKHET
NAINITAL
KATHGODAM
HALDWANI
TANAKPUR
KOTDWAR
DHAMPUR
BIJNOR
HARDWAR
DEHRADUN
GANGA
LEGEND
ALLUVIAL PROVINCE
SIWALIK PROVINCE
CALCAREOUS PROVINCE
OF SCHISTOSE PROVINCE
CRYSTALLINE PROVINCE
GLACIAL PROVINCE

Fig 1

Similar type of springs are also observed in U.P. Himalaya at different part in the cale facies.

6. Granite and Gneisse Province (Crystalline-Province)

These rocks occupies about 8000 km^2 in the part of Garhwal/Kumaun Himalaya. It contains water only along fracture zones or in the valley fills, and in the weathered portion of granites. In the Purola area the large fanglomerated may also be tried for ground water and may be able to support the tube wells of moderate yields. The higher altitude terrain above 8000' ft. specially in rural areas yields in fair quantity which can be supplied through pipe by gravity flows.

7. Glacial Province

In the glacial areas the melting yield water which is through in to the various channels, flows in to main river, but generally there is no habitation or it is very sparsh. The areas which are located at more than 8000 ft. height in the Himalayan part containing the rock with such water holding capacity, even as snow peak glaciers, morainic deposits. These morainic material act as a good aquifers. These terrain is having a large number of Tals, lakes of glacier origin and springs too. The water of these tals, lakes can be used through pipelines for drinking and irrigation purposes.

Water Cycle

In Himalayan region, the maximum amount of water is stored periodically by rainfall in the form of surface water and also for the Ground water. The water constantly evaporates from all open water surface, including sea, reservoir or wet soil. The vegetation on earth's surface or even the biological world is general arguments evaporation in the form of transpiration. The water vapour accumulates into clouds which gives rise to rain.

All the rainfall on surface of earth distributed as runoff evaporation and infiltration. The runoff i.e. water flows along the streams a rivers vegetation soil loisture are spent through evaporation or by way if being released to the sea. The infiltration water is divided into two parts, firstly it is hold on surface of the earth and secondly it flow as ground water in the deeper regions of earth's surface.

Ground water is released to the surface by natural springs or it is takenout by man made actions. In both the cases the water when released on the surface of earth either as runoff or evaporation.

The runoff through melting of ice in summers in the Hilly areas plays important role for the perennial rivers, i.e. the Ganga, the Yamuna, the Kali and the Ram Ganga rivers.

The water held in soil is passed to the deeper rock masses. This water occupies all the open areas within the rocks and flow according the sub surface ground slope conditions. It can be imagined that like the surface streams of the watershed ground water will have to same direction of flow. It cannot however be measured as to how much ground water is exactly flowing out. An approximate distribution of water, various accounts is composed in Table as guide to compute the water balance.

Water Harvesting

The mountainous areas of U.P. spread over 51125 sq. km. along the Himalaya with a population of 48.36 Lakhs (1981 census) are blessed with rich forest, Ganga, Yamuna and their tributaries and a labytinth of small streams. U.P. hills have a large share of water resources then any other part of the country.

The term Water Harvesting was first defined as the collection and storage of rain water for its use in irrigation. Recently it used as the process of collecting natural precipitation from watershed for beneficial use. The improved crop varieties certainly give more yield but they consume more water too. If the quality of stored water is improved, it can be well consumed by humans and river stock. The conservation of water can be done by following.

(1) Through construction of artificial reservoirs by making barrages and small dams on the various river at suitable sites. Therefore, the more amount of surface water will be percolated and added to the ground water in lower ridges for the supply of drinking water. The stored water can be used for the irrigation through canal and pipelines.

(2) Through the Hydrums technology : The water powered water lifting device which already in operation in U.P. hills. Hydrums are most appropriate in the context of interior villages for both community drinking water and also irrigation schemes. The Hydromechanical power unit (Hydrum) can be run on a cooperative basis the skills necessary to operate and maintain hydro-mechanical units can be transferred to the village mechanics.

(3) Through Micro Hydro Schemes: In view of the side effects of such major products decentralised power generation projects, we must preferred micro hydel projects. After utilisation of water of rivers/streams from electricity generation it can be utilised for drinking and irrigation purposes through pipe lines and some amount of water can be stored in the cement concrete tank for future utilizations etc. Also it has been experienced that the water stored in closed tanks presents any infection by bacteria. The tank should be made of local materials to a great extent in order to reduce burden on the farmers. This system would drudgery on farm women and improve their purchasing power.

(4) Through rotary drive hand-pump/rover pump: The load carrying capacity of hill peoples are tremendous. In some of the valleys where ground water level are shallow rotary drive hand pump and rover pumps can be tried for small scale irrigation and drinking water.

(5) Through collection of Rain Water: The construction of small bunds along streams, stream let dry nalas for catching rain water. The water from the dry nalas in rain should be lead into cement concrete tank to be filtered, treated and stored in the tanks for domestic and drinking purposes. The rain water should be collected through the roofs of houses and stored in tank, then supplied through pipelines and with the help of tullu pump it can be lifted upto the height of 50 ft.

In the areas where the smaller habitations live the lifting of water may be economical. It is suggested that rain water may be either collected from the runoff from the roof of the house or from dry nalas during monsoon. The large storage under ground tanks with or with catching system, after an inexpensive treatment, it may used for domestic consumptions.

Conclusion

(1) The ground water investigations will be undertaken in the areas of water scarcity.

(2) The systematic pipe line will be laid to harness the water resources of the area, altitudes at much above than the main stream levels, utilizing through gravity flow.

(3) Even if turn it to be initially expensive project, in the long run it will be cheaper than the lift irrigation.

REFERENCES

Anil Kumar, 1988, Land Management and Water Harvesting. Proceeding Water Manage in U.P. Hill 35-38.

Arya, K.L., 1988, Parvatiy Kshetra Me Samaikit Jalagam Parbandh Pariyojna - 55-59.

Auden, J.B., 1937, Structure of the Himalaya in Garhwal Rec. Geologic Survey, India, 71, (1937).

Arya, K.L., 1988, Parvatiya Kshetra Me Samaikit Jalagam Prabandh Pariyojana Proc. Water Manage in U.P. Hills.

Fouzdar, D., 1987, Water Resources. Proc. Water Management, U.P. Hills; 4-16.

Gary, S. P., - Principles of Ground Water Geology, Publication New - Delhi.

Jain, J.K., 1977, "India Under ground Water Resources" Phil - Trans - Royal Socio. India.

Kala, G.P., 1988, "Madhya Himalaya : Jal Prabandh Yojana." Proc. Water Management in U.P. Hills.

Nautiyal, S.P., 1943, Artesian Wells of Jonsar Bhabar Area Geol. S. I. (1950)

Shandilya, A.K., 1987, High altitudinal lakes & Tourism in Garhwal Himalaya.

M.S. Sati and Kanduri, Y.S., Proceeding Inst. on Impact of Tourism on Mount Environment G.U. Srinagar March, p - 89.

Shankar, Uday C., 1988, Water & Energy Proceeding Water Management in U.P. Hills.

Singh J.B., 1988, Water & Land, "Water Management in U.P. Hills; pp-17-20.

Todd, D.K., 1959, Ground Water Hydrology. John Willky And Sous. New York.

Valdiya, K.S., 1980, Stratigraphy and Correlation of cesser Himalaya. Hindustan Publishing Corp. New Delhi.

Varma, J.L., 1988, Pauri Ka Parbandh Proc. Water Management in U.P. Hills.

2

STRUCTURE AND DIVERSITY OF A MONTANE FOREST IN A PART OF BHAGIRATHI VALLEY, GARHWAL HIMALAYA

G.S. Rajwar

Four oak forest sites (two east facing and two west facing) above the village Manpur in Uttarkashi district of Garhwal Himalaya were studied for structure, phytosociology and diversity. Upper slopes on both the aspects comprised undisturbed and the lower slopes represented slightly disturbed sites. These oak forests are mainly composed of the 'banj' oak *(Quercus leucotrichophora)*. Lower slopes on both the aspects consisted of oak-pine mixed forest *(Quercus leucotrichophora-Pinus roxburghii)*. Upper slope sites on the east and west facing aspects represented *Quercus leucotrichophora* and *Quercus floribunda* - *Q. leucotrichophora* forests respectively. The total density values of trees ranged from 1020 to 2460 trees ha^{-1} and the range of basal cover values was 46.17 to 71.23 $m^2 ha^{-1}$. These values were higher than those for other montane and temperate forests. The structure and diversity of these forests have been compared with other montane forests of Himalaya and temperate forests of the world.

Introduction

Oak forests are usually distributed at elevations between 1200 and 2450 m in the Western Himalaya. Lower elevations in this range are occupied by oak-pine mixed forest. In Kumaun and Garhwal Himalaya this type of forest is represented by *Quercus leucotrichophora* and *Pinus roxburghii.* Other species of oak are found in different altitudinal zones of Garhwal Himalaya above the oak-pine mixed forest (Osmaston, 1922). The climax species of oak are *Quercus leucotrichophora* at low altitudes and *Q. semecarpifolia* at high altitudes (Troup, 1921). Montane forests are usually distributed between submontane chir pine forests and subalpine forests in the Himalaya. *Quercus leucotrichophora* and *Q. floribunda* are two important species of oak in the Himalayan region. These are used as fuel, fodder and in the manufacture of agricultural implements. The oak forests are highly correlated with the natural springs in the Himalayan region. There are many sources of springs in these forests as compared to the forests up and down these forests. Heavy biotic stress is posing a great threat to these forests in the recent years.

Many studies have been conducted on the structure, diversity and regeneration of oak and oak-conifer mixed forests of Garhwal and Kumaun Himalaya. There are reports on the studies of oak forests of Garhwal Himalaya since the third decade of the present century (Osmaston, 1922; Dudgeon and Kenoyer, 1925). Studies on the structure, succession and the impact of biotic stress and regeneration of oak and oak-pine mixed forests of Garhwal Himalaya have been published during the past 30 years (Gupta and Singh, 1962; Gupta, 1966; Joshi *et al.*, 1985; Rajwar, 1988, 1989; Puri *et al.* 1989; Rajwar and Dobhal, 1990). Ecology of *Pinus roxburghii* has been studied by Joshi and Joshi (1982) and Rajwar (1988). Different ecological studies on the oak and oak-pine mixed forests of Kumaun Himalaya have also been published (Saxena *et al.*, 1978; Ralhan *et al.*, 1982; Upreti *et al*, 1985). A little attempt has been made to study the diversity of oak and oak-pine mixed forests of Garhwal Himalaya.

In the present paper four oak forest sites, two each on east facing and west facing slopes above the village Manpur in Bhagirathi Valley in Uttarkashi district of Garhwal Himalaya have been described for their structure and diversity.

The Study Area

The study sites are located at 30°N latitude and 78° E longitude above the village Manpur in the district Uttarkashi in Bhagirathi Valley of Garhwal Himalaya. Village Manpur is situated at a distance of 8 Km from Uttarkashi city in the north-east direction. Four slope sites were selected for the study, two each on east facing and west facing slopes. The altitude varied from 1220 to 2300 m. The upper slope sites were located above the elevation of 1950 m in this range. East facing slopes had more slope angle (28° - 40°) than that on the west facing slopes (20° -35°). Upper study sites represented undisturbed sites on both the slopes whereas lower slope sites has a low degree of disturbance. The causes of disturbance included partly lopping of trees near the village, destruction of shrubs, trees and their saplings for fuel and fodder and by grazing and browsing by the domestic animals. There were fires from April to June on the lower slopes.

The climate of the study area is monsoonic montane temperate. There are three distinct seasons in a year: summer (April to mid-June), rainy (mid-June to September) and winter (October to March). Frosts and snowfall are common during the period December-February in the upper slope forests. The total rainfall for the city Uttarkashi for the study period February 1985 to January 1986 was 1674 mm. June was the hottest month (2. 5°C) was recorded for the month January. The study area belongs to lower Himalayan Tertiary formations geologically. The soils are derived from quartzite, shale, granite and limestone rocks.

Methodology

The individuals of tree species were sampled for the study according to the classification of Mueller-Dombois and Ellenberg (1974). The tree and shrub layers were analysed by quadrat method using the sizes 10 x 10 m and 5 x 5 m respectively. Phytosociological data were analysed following the methods of Curtis and Mc Intosh (1950). The index of dominance or the concentration of dominance (C) was calculated using Simpson's (1949) index as:

$$C = \Sigma (n_i/N)^2$$

where n_i is the importance value for the species i, and N is the total importance value of all the species in a stand. The index of diversity ($\overline{H}$) was computed using Shannon and Weaver's information index (Shannon and Weaver, 1963) as:

$$\bar{H} = \sum_{i=1}^{S} (n_i/N) \log (n_i/N)$$

In the present study, C and $\bar{H}$ were calculated on the basis of density values.

Results

Structure of Vegetation

Tree Layer

Basal cover and density values of dominant tree species on the four slope sites have been given in Table - 1. The number of tree species varied from 10 (lower east facing slope) to 16 (upper west facing slope).

TABLE 1: **Density and Basal Cover Values of Dominant Tree and Shrub Species for Four Slope Sites of a Montane Forest in Bhagirathi Valley.**

Species	*Basal Cover m^2 ha^{-1}*	*Density plants ha^{-1}*
1	2	3
UPPER EAST FACING SLOPE (*Quercus leucotrichophora* forest)		
TREES		
Quercus leucotrichophora	23.52	380
Rhododendron arboreum	9.26	360
Lyonia ovalifolia	8.28	300
Quercus floribunda	6.14	180
Persea duthiei	2.12	120
SHRUBS		
Berberis chitria	3.62	1280
Zanthoxylum alatum	1.41	1160
Indigofera heterantha	1.21	980
LOWER EAST FACING SLOPE (*Quercus leucotrichophora - Pinus roxburghii* forest)		
TREES		
Quercus leucotrichophora	20.42	360
Pinus roxburghii	15.12	180
Rhododendron arboreum	16.84	200
Lyonia ovalifolia	4.35	120
Pyrus pashia	2.25	100
Alnus nepalensis	1.18	80
SHRUBS		
Berberis chitria	3.81	1120
Woodfordia fruticosa	1.16	1110
Rubus ellipticus	0.84	1080
Zanthoxylum alatum	0.28	1160
Murraya koenigii	0.07	620
UPPER WEST FACING SLOPE (*Quercus floribunda-Q. leucotrichophora* forest)		
TREES		
Quercus floribunda	22.82	380
Q. leucotrichophora	18.52	320

Contd.

1	2	3
Rhododendron arboreum	10.84	300
Viburnum cotinifolium	4.82	280
Cornus capitata	2.62	120
Persea duthiei	1.18	80
SHRUBS		
Berberis chitria	5.12	1320
Rubus niveus	3.84	1160
LOWER WEST FACING SLOPE (*Quercus leucotrichophora-Pinus roxburghii* forest)		
TREES		
Quercus leucotrichophora	18.02	280
Pinus roxburghii	15.46	220
Rhododendron arboreum	7.25	200
Pyrus pashia	4.12	120
SHRUBS		
Rubus elliptica	3.81	1260
Berberis chitria	2.08	1220
Rosa moschata	1.81	880

On the east facing slope, the upper site had the higher values of total basal cover (62.52 m^2 ha^{-1}) and density (2080 trees ha^{-1}). *Quercus leucotrichophora* dominated both the east facing slopes with maximum values of the density and basal cover (Table-1). Other trees in the order of importance on the upper east facing slope were *Rhododendron arboreum, Lyonia ovalifolia, Quercus floribunda* and *Persea duthiei*. On the lower east facing slope site *Pinus roxburghii* was the subdominant species. Other important species included *Rhododendron arboreum, Lyonia ovalifolia, Pyrus pashia* and *Alnus nepalensis*. On the basis of phytosociological values, the forests on upper and lower east facing slopes may be designated as *Quercus leucotrichophora* and *Quercus leucotrichophora - Pinus roxburghii* forests respectively.

Total basal cover and density values of trees were also higher for the upper slope on the west facing aspect (71. 23 m^2 ha^{-1} and 2460 trees ha^{-1} respectively). The upper slope site on the west facing aspect was dominated by *Quercus floribunda* showing highest basal cover (22.82 m^2 ha^{-1}) and density values (380 trees ha^{-1}). This slope forest was subdominated by *Quercus leucotrichophora* (Table-1). Other important tree species on the upper west facing slope were *Rhododendron arboreum, Viburnum cotinifolium, Cornus capitata* and *Persea duthiei*. The lower west facing slope was dominated by *Quercus leucotrichophora. Pinus roxburghii* subdominated this forest site Table-1. Other important trees on this site were *Rhododendron arboreum* and *Pyrus pashia*. On the basis of phytosociological data, the forest types on the upper and lower west facing slopes may be designated as *Quercus floribunda - Q. leucotrichophora* and *Quercus leucotrichophora - Pinus roxburghii* respectively.

Shrubs

The number of shrub species varied from 9 to 16 in these oak forests. The west facing slopes recorded more shrub species than by the east facing slopes. The values of density and basal cover of dominant shrub species have been given in Table-1. *Berberis chitria* had the maximum values of basal cover (5.12 m^2 ha^{-1}) and density (1320 plants ha^{-1}) on the upper west facing slope. Other dominant shrub species included *Rubus niveus, Zanthoxylum alatum, Indigofera heterantha* on the upper slopes. The lower slopes were dominated by *Rubus elliptica, Berberis chitria, Rosa moschata, Wood fordia fruticosa, zanthoxylum alatum* and *Murraya koenigii.* The highest total basal cover (16.81 $m^2\ ha^{-1}$) and density values (10860 plants ha^{-1}) were shown by the upper west facing slope and the minimum values were exhibited by the lower east facing slope.

TABLE 2: **Total Density and Basal Cover Values of Trees and Shrubs for the Slope Sites of a Montane Forest in Bhagirathi Valley**

Slope site	*Forest*	*Total Density plants ha^{-1}*		*Total Basal Cover $m^2\ ha^{-1}$*	
		TREES	*SHRUBS*	*TREES*	*SHRUBS*
Upper East Facing	*Quercus leucotrichophora*	2080	6840	62.52	10.08
Lower East Facing	*Quercus leucotrichophora-Pinus roxburghii*	1020	5980	46.17	6.29
Upper West Facing	*Quercus floribunda - Q. leucotrichophora*	2460	10860	71.23	16.81
Lower West Facing	*Quercus leucotrichophora-Pinus roxburghii*	1040	8480	52.18	11.27

Indices of Dominance and Diversity

The index of dominance (concentration of dominance) values for trees and shrubs of the present montane forests ranged from 0.09 to 0.23 and 0.09 to 0.21 respectively, Table-3.

TABLE 3: **Index of Dominance (C) and Index of Diversity ($\bar{H}$) Values for Trees and Shrubs for the Slope Sites of a Montane Forest in Bhagirathi Valley**

Slope Forest		C		$\bar{H}$	
		Trees	*Shrubs*	*Trees*	*Shrubs*
Upper East Facing	*(Quercus leucotrichophora)*	0.12	0.13	1.19	1.22
Lower East Facing	*(Quercus leucotrichophora-Pinus roxburghii)*	0.09	0.09	0.88	0.70
Upper West Facing	*(Quercus Floribunda-Q. leucotrichophora)*	0.23	0.21	1.28	1.31
Lower West Facing	*(Quercus leucotrichophora-Pinus roxburghii)*	0.21	0.11	0.76	1.01

The diversity index for trees varied from 0.76 to 1.28. Shrubs showed a range of diversity index from 0.70 to 1.31. Maximum values of the concentration of dominance and diversity index were shown by the upper west facing slope forest.

Discussion

Oak forests of the present study area were dominated by *Quercus leucotrichophora* on three slope sites on the basis of basal cover and density. The upper west facing slope exhibited *Quercus floribunda - Q. leucotrichophora* community, whereas, the lower slopes of both the aspects (disturbed sites) represented oak-pine mixed *(Quercus leucotrichophora - Pinus roxburghii)* forest. The number of species of trees varied from 10 to 16 and of shrubs this range was 9 to 16.

The total tree density values ranged from 1020 to 2460 trees ha^{-1} which falls within the range of values (350 to 2840 trees ha^{-1}) for other montane and temperate forests of the world (Duvigneaud and Denaeyer De-Smet, 1970; Dabel and Day, 1977; Crow, 1978; Killingbeck and Wali, 1978; Saxena and Singh, 1982). The total tree basal cover values (46.17 to 71.23 m^2 ha^{-1}) for the forests of the present study are mostly toward the upper limit of the total basal cover range (12.2 to 83.8 m^2 ha^{-1}) for other montane and temperate forests of the world (Duvigneaud and Denaeyer De-Smet, 1970; Reiners, 1972; Dabel and Day, 1977; Crow, 1978; Killingbeck and Wali, 1978; Saxena and Singh, 1982; Ralhan *et al.*, 1982; Upreti *et al.*, 1985; Reiners and Lang, 1987; Rajwar and Dobhal, 1990), but the undistributed upper slope forests had more values of density and basal cover. The lower slopes consisting of oak-pine mixed forests showed a low degree of biotic interference. The number of partly-lopped and half-lopped trees was less than 20% on these slopes.

The diversity index ($\overline{H}$) range for trees and shrubs (0.76 to 1.28) falls within the range of values of $\overline{H}$ (0.5 to 3.4) for different montane and temperate forests of the world (Monk, 1967; Risser and Rice, 1971; Saxena and Singh, 1982; Ralhan *et al.*, 1982; Upreti *et al.*, 1985; Rajwar, 1989, Rajwar and Dobhal, 1990). More diversity was reflected by the upper slope forests. The west facing aspect favoured more density, basal cover, dominance and diversity. Thus, oak forests of the present study area represent a conducive status of ecology and a high volume of biomass. The west facing aspect was more favourable for regeneration and better performance of trees and shrubs.

REFERENCES

Crow, T.R., 1978. Biomass and production in three contiguous forests in northern Wisconsin. *Ecology* 54: 266-273.

Curtis, J.T. and R.P. Mc Intosh, 1950. The interrelation of certain analytic and synthetic phytosociol ogical characters. *Ecology* 31: 434-455.

Dabel, C.V. and F.P. Day, 1977. Structural comparisons of four plant communities in the Great Dismal Swamp, Virginia. *Bull. Torrey Bot. Club 104:* 352-360.

Dudgeon, W. and L. A. Kenoyer, 1925. The ecology of Tehri Garhwal, a contribution to the ecology of the Western Himalayas. *J. Ind. Bot. Soc.* 4: 233-284.

Duvigneaud, P. and S. Denaeyer De-Smet, 1970. Biological cycling of minerals in temperate deciduous forests. In: Reichle, D.E. (Ed.), *An Analysis of Temperate Forest Ecosystems.* Springer-Verlag, Berlin-*Hei delberg* -New York.

Gupta, R.K., 1966. Studies in the succession of the oak-conifer forests of Garhwal Himalaya. *Trop. Ecol.* 7: 67-83

Gupta, R.K. and J.S. Singh, 1962. Succession of vegetation types in Tons Valley of the Garhwal Himalayas. *Ind. For.* 88: 290-295.

Joshi, S. and Anil P. Joshi, 1982. Ecology of *Pinus roxburghii* in Western Himalayas. *Himal. J. Sci.* 2: 41-44.

Joshi, S.P., A. Raizada and M.M. Srivastava, 1985. Vegetational analysis of oak forest at Mussoorie. *Ind. J. For.* 8(1): 78-84.

Killingbeck, K.T. and M.K. Wali, 1978. Analysis of a North Dakota Gallery forest: nutrient, trace element and production. *Oikos 30:* 29-60.

Monk, C.D., 1967. Tree species diversity in the eastern deciduous forest with particular reference to north central Florida. *Amer. Nat. 101*: 173-187.

Mueller-Dombois, D. and H. Ellenberg, 1974. *Aims and Methods of Vegetation Ecology*. John Wiley & Sons, New York.

Osmaston, A.E., 1922. Notes on the forest communities of the Garhwal Himalaya. *J. Ecol.* 10: 129-167.

Puri, G.S., R.K. Gupta, V.M. Meher-Homji and S. Puri, 1989. *Forest Ecology,* Vol. 2. Oxford & IBH Publishing Co. Pvt. Ltd., New Delhi.

Rajwar, G.S., 1988. The forest vegetation on the east and west facing slopes of the Bhagirathi Valley near Uttarkashi, Garhwal Himalaya. *In* : Prakash, R., S.S. Negi and M.P. Shiva (Eds.), *Advances in Forestry Research in India,* Vol. 1, International Book Distributors, Dehra Dun, pp.199-206.

Rajwar, G.S., 1989. Human impact on the forests of Garhwal Himalayas. In : Prakash, R. (Ed.), *Advances in Forestry Research in India,* Vol. 3. International Book Distributors, Dehra Dun, pp. 229-241.

Rajwar, G.S. and L.P. Dobhal, 1990. Ecological studies in some oak forests of Western Garhwal Himalaya. *In* : Rajwar, G.S. (Ed.), *Advances in Himalayan Ecology*. Today and Tomorrow's Printers and Publishers, New Delhi (in press).

Ralhan, P.K., A.K. Saxena and J.S. Singh, 1982. Analysis of forest vegetation at and around Nainital in Kumaun Himalaya. *Proc. Ind. Natl. Sci. Acad. 48:* 122-138.

Reiners, W.A., 1972. Structure and energetics of three Minnesota forests. *Ecol. Monogr.* 42: 71-94..

Reiners, W.A., and G.E. Lang, 1987. Changes in litterfall along a gradient in altitude. *J. Ecol. 75:* 629-638.

Risser, P.G. and E.L. Rice, 1971. Diversity in tree species in Oklahoma upland forests. *Ecology 52:* 876-880.

Saxena, A.K. and J.S. Singh, 1982. A phytosociological analysis of woody species in forest communities of a part of Kumaun Himalaya. *Vegetatio 50*: 3-22.

Saxena, A.K., U. Pandey and J.S. Singh, 1978. On the ecology of oak forests in Nainital hills, Kumaun Himalaya. *In* : Singh, J.S. and B. Gopal (Eds.), *Glimpses of Ecology*. International Scientific Publications, Jaipur, pp. 167-180.

Shannon, C.Z. and W. Weaver, 1963. *The Mathematical Theory of Communication*. University of Illinois Press, Urbana.

Simpson, E.H., 1949. Measurement of diversity. *Nature 163*: 688.

Troup, R.S., 1921. *The Silviculture of Indian Trees,* Vols. I-III. Clarendon Press, Oxford.

Upreti, N., J.C. Tewari and S.P. Singh, 1985. The oak forests of Kumaun Himalaya (India) : composition, diversity and regeneration. *Mount. Res. Dev.* 5(2): 163-174.

3

GEO-ECOLOGY AND FOREST DEVELOPMENT OF U.P. HIMALAYA

D.D. Maithani

Introduction

The Himalayas are the youngest mountains in the world. They provide a variety of natural resources to the Indian sub-continent, including the life-giving water. In order to Himalayan Geo-ecology is one of the most important and most threatened of life support system on earth. In the shadow of Himalayas live more than 155 million people in India, some of them poorest in the world. The population pressure on land and its mounting demand for fuel and fodder has already denuded areas of the Himalayan ranges. In U.P. Himalaya the rate of deforestation is much faster than the rate of afforestation. Structurally, the Siwaliks are made up of fresh water deposits composed largely of clays, sandstones and conglomerates. These rocks show only variable sediments during heavy rains. The zonal distribution of forests from the outer hills through the central ranges is determined primarily by altitude, soils, climate and other biotic factors. Altitude determines subtropical temperate and alpine forests from the sub-montane tracts to the snowy ranges.

In U.P. Himalaya, there are two transverse zones. They are:

(a) Western U.P. Himalaya or Garhwal Himalaya.

(b) Eastern U.P. Himalaya or Kumaon Himalaya.

The Garhwal Himalaya comprises five districts, viz. Uttarkashi, Tehri, Pauri, Chamoli and Dehradun while the Kumaon Himalaya has three districts, viz. Pithoragarh, Almora and Nainital. (Table 1) with the total Geographical Area of 51,033 Km^2.

Geo-ecology and Forest Cover

Geo-ecology is essential to the wise management of our earth. The Geo-ecological challenge is to preserve the diversity of the earth's living systems. In the past years have seen immense advances in our understanding of ecology, especially through such initiatives as the international Biological programme and its successor, Unesco's Man and Biosphere (MAB) Programme. This research has illuminated some of the intricate relationship between living organisms and between life and its physical setting.

Understanding of Geo-ecology, forest eco-system contribute a lot of organic matter of the soil in the shape of leaf fall, fruits, branches which influence soil properties in many was apart from being sources of nutrients of the plants, the nature and amount of this contribution depending, besides other- factors on the species present.

In view of U.P. Himalaya, forests are recognised to be powerful Geo-ecological unit affecting environmental besides being the most important renewable and natural resources. In Kumaon Himalaya, Pithoragarh district had maximum per cent land under snow cover, high altitude meadows and non-forest use. Almora and Nainital districts also occupied high per cent area in non forest uses. The forested land was 1785 Km^2 in Almora district, of which was under medium, poor and good forests accounted for 63.8%, 22.6% and 13.4% respectively (Table 1). Similarly in Nainital district, medium forests were found in 62.5% and Pithoragarh district in 56.9% of the forested land.

In Garhwal Himalaya, Pauri and Dehradun districts, occupied 64.2% and 76.7% area under non-forested land-use. Snow cover and high altitude meadows were absent in these districts. The area of Uttarkashi district is 3755 Km^2, which was mainly under snow cover and high altitude meadows.

The per cent of total forest covered area was higher in Pauri and Tehri districts compared to other districts of Garhwal region. In Pauri district, 'medium' forests were found in 49.9% and Tehri district 21.2% of the total forested land. In Uttarkashi district, forested land accounted for 43.9% area of medium forests. Through Chamoli and Dehradun districts, occupied 63.0% and 28.7% of medium forests of the forested land, respectively (Table 2). In U.P. Himalaya, total 28.6% of the area was under forests. 'Good' forests accounted for only 15.5% area of the forested land.

TABLE 1: **Forested Landuse in U.P. Himalaya (Area Km^2)**

Landuse	*Garhwal Himalaya*					*Kumaon Himalaya*		
Forests	*Dehra Dun*	*Pauri*	*Uttar-kashi*	*Tehri*	*Cha-moli*	*Pittho-ragarh*	*Almora*	*Nainital*
(i) Good Forest	260	152	330	442	70	524	240	255
(ii) Medium Forest	207	973	490	360	1262	1340	1140	1890
(iii) Poor Forest	252	825	294	898	670	490	405	875
Total Forest	719	1950	1114	1700	2002	2354	1785	3020

Source: Working Plans of the Districts and Forest Department.

TABLE - 2: **Forest Area in U.P. Himalaya**

	Garhwal			*Kumaon*		
Forests	*Area (Km^2)*	*% of Forest Area*	*%of Total Area*	*Area (km^2)*	*%of Forest Area*	*%of Total Area*
1. Good Forest	1254	16.75	4.18	1019	14.23	4.84
2. Medium Forest	3292	43.98	10.97	4370	61.04	20.77
3. Poor Forest	2939	39.27	9.79	1770	24.73	8.41
Total	7485	100.00	24.95	7159	100.00	34.03

Source: Working Plans of the Districts and Forest Department.

Forest constitute an enormous wealth of the region. The principal component of the forest below the elevation of 1500 m is Sal (Shorea-robusta), Semal (Salmafia-malabarica), Khair (Acacia-catechu), Haldu (Aldina-cordifolia) and Pine, which is sub-tropical forest zone. Temperate forest zone (about 1500 m - 2000 m) generally consist of Pine, Oak, Deodar and some species of deciduous forests. Silver fir, Banj Oak, Maru Oak, Blue pine, Deodar and Birch etc. constitute sub-alpine forest zone (about 2000 m - 3000 m.), while on high Himalaya (about 3000 m.) contains alpine posture locally known as 'Bugyal' with a rich variety of Polygonum, Swertia, Salix, Epilobium and Grasses.

Forest Denudation and Destruction

Deforestation caused by the need for domestic fuel, timber supplies and livestock fodder, over grazing of natural grasses, indiscriminate cultivation of steep slopes combined with poor agricultural practices and badly designed roads, all contribute to general Geo-ecological decline observed in many parts of U.P. Himalaya. The outcome of these activities is increasing soil erosion with consequent damages.

Over the last two decades there has been a rapid and intensive development of U.P. Himalayan region, top priority was given to connect these areas with the plains by constructing main motor road, feeder and link roads, The impact of human activities, has led to serious Geo-ecological repercussions which have been caused mainly as below:-

(i) Due to motor roads occurrence of landslides and erosion on mass scale. According to Dr. K.S. Valdiya, the construction of 44,000 Km. motor roads in the Himalaya has generated 2650 million cubic meters of debris.

(ii) Increase in population and the diverse pattern of land utilization, resulting in large scale destruction of forests along the hill slopes. Some areas of Ramganga RF., Alaknanda RF., Bhagirathi RF., Ganga (B) and Upper Yamuna RF. are the examples.

(iii) Growth of industries and Mining on all sides, such as Mussoorie hills, Doon valley, Jhiroli (Almora), Chandhak (Pithoragarh) and Mohankhal (Chamoli) are the examples of the irreparable losses occurred due to the destructive system of quarrying.

(iv) The pressure of livestock population on grazing lands and migratory graziers are a potent force responsible for the continual degradation of forest resources. Bhotias, Gaddis, Khadwals and Gujars are the main graziers in the region. This clearing of forests has caused severe erosion and land degradation.

(v) Construction of dams over the hill rivers, occur the destruction of forests in the valley-slopes, floods on fertile land and the problems of human rehabilitation. The number of such projects over Ganga and its tributaries in the hill are twenty two (S.L. Bahuguna, 1985) one of such projects is Tehri Dam over Bhagirathi river in U.P. is under construction.

Forest Geo-ecology in U.P. Himalaya have been providing a variety of forest products and services for long past. But the rising demands of these products goods and services due to rapidly increasing livestock and human population, agricultural and land for urban uses coupled with the low productivity of our forests has caused a very critical situation. Much of the forests in the region are situated in the sensitive catchment areas of important rivers and maintenance of a forest cover in these catchment areas is very necessary.

Forest Conservation and Development

The study indicated the poverty of the forest vegetation in the U.P. Himalaya. Increased human population and expanding agriculture are the principal causes of forest deterioration (Singh et., 1984). The National Forest Policy, 1952, has specifically mentioned that the forests in these ecologically crucial areas should be managed as protection forests. In the meeting of the International Union for Conservation of Nature's Commission for National Parks and protected areas, held at Corbett National Park, it was noted that in spite of the efforts of governmental and other agencies, certain regions are still endangered by ill planned development, encroachment, illegal felling and mismanagement of resources. India has 53 National Parks and 247 sanctuaries and the annual forest loss has been reduced to about 4,000 ha. from 1.5 Lakh ha. in 1980 (Meher Homji, The Illustrated Weekly of India, 22.6. 1986, p. 21).

In this context of increasing productivity from our forest rehabilitation of degraded forest lands caused by large scale destruction of tree cover in the U.P. Himalayan region. 'Chipko' movement of Garhwal could be treated as the fore-runner of social forestry movement in U.P. Himalaya and India as well. In the words of Sunder Lal Bahuguna, 'Every standing green tree in the Himalaya is a sentry guarding against the fury of the floods and landslides, as it performs the important duty of conserving rain water in the hills'. Now, the Himalayas are in critical stage of deterioration and only a concerted and integrated action by all concerned is urgently needed. The following suggestions are given below:

(A) Formation of local conservation and forest societies in U.P. Himalayas, with the several programmes such as Eco-development camps, Nurseries and Plantation, should be promoted.

(B) It is urgent that people's participation must be promoted for such a Himalayan Geo-ecological programme. Fuel and fodder are the most acute problem faced by women. Hence, women's participation is essential in any agro-forestry programme. If they can be given some income earning opportunities during their spare time, the pressure on forests can be reduced to some extant.

(C) Trees should be planted on field risers for fuel and fodder, the technique of farm forestry in open grazing areas should be developed which is lacking at present in the region.

(D) Need for development of bamboos in hill areas and Cane (Ringaal) in the North-eastern slopes to develope the samll scale industries.

(E) Geo-ecology and development must be fully integrated so that the needs of people, cattle and environmental are not met.

(F) Geo-ecosystem concept and environmental education must be introduced right from the primary school level with emphases to field studies and trips.

The overgrazing, soil erosion, droughts and forest fires have further deteriorated the forest wealth and this is an unending process. The check the further damage of the flora, it is necessary that some measures are taken right now.

The forest policy for the Himalayan region has been discussed at several regional, national and international levels (Tiwari 1982, Shah et. 1984, Pande 1984, Joshi 1985). It should aim at balancing the productive and protective aspects. All forest programmes must have a scientific working plan aiming to provide sustained yield for long-run. The transport and marketing aspects of the farm or social forestry projects require serious considerations. Existing forest land, irrespective of its legal status and ownership, should not be alienated for other landuses, particularly agriculture and horticulture. Studies on the silvipastoral system and tree-grass associations in different agro-climatic conditions must be needed. In high altitude areas where climatic conditions do not favour of tree growth, extensive summer pastures have given rise to sheep and cattle rearing. There it is primarily an animal-based economy as they provide food, wool, milk and also means of transportation. Local wisdom and traditions should be given top priority.

REFERENCES

Anon. (1982). Study Report on the Socio-economic Development, of the Himalayan Hills, Planning Commission, Govt. of India, p.132.

Davis, C.G. Higging (1982). Grazing-step Terracettes and Their Significance. Z. Geomosph. 26(4): 459-472.

Joshi, D.P. (1985). 'Forest Policy for the Regeneration of the Himalayan Environments'. In J.S. Singh (Ed.) Environmental Regeneration in Himalaya Concept & Strategies, CHEA & Gyanodaya Prakashan, Nainital, pp. 260-285.

Puri, K.S. (1986), 'Water-shed Management for Environmental Security in Himalayan Region'.

Dewan, M.L. & S. Sharma (1985) : 'People's participation in Himalayan Eco-system Development'. Centre for Policy Research.

Maithani, D.D. (1983), Unpublished D. Phil. Thesis, 'Tehri Garhwal — A Geographical Appraisal of Resources'. Garhwal University, Srinagar.

Singh, J.S. ; U. Pandey & Tiwari (1984), Man & Forests: A Central Himalayan Case Study. Ambio, 13; 80-87.

Shah, S.L. (1982). 'Ecological Degradation & Future of Agriculture in the Himalayas' Indian J. Agr. Econ. 37; 1-22.

Wadia, D.N. (1919), Geology of India, Tata McGraw Hill Pub. Co. New Delhi, IV edition (Reprinted, 1978), p. 508.

4

GEOMORPHOLOGICAL INVESTIGATIONS IN THE NORTH-EAST OF SRINAGAR, GARHWAL HIMALAYA

A. K. Sandilya and R.S. Negi

A detailed geological and geomorphological investigations were carried out in the north-east part of the Srinagar Garhwal area in the lesser Himalaya. (This area consists of the rocks of Chandpur Groups Schistose and slaty phyllites), arenaceous and argillocalcareous rocks of the Calc Group and gneisses and schists of the Crystalling Group Tectonically these rocks are bounded by the north Almora Thrust in the south, Pratapnagar Thrust in the central part and Main Central Thrust in the northern part of the area.

The geomorphic features were studied and their relationship with these thrusts has been established. It is found that the rocks of the Chandpur and Calc Groups are more sensitive to the deformation than that of the Crystalline group thus the landforms developed are fully controlled by the tectonic alignments. In the northern part of the study area the rocks of the Crystalline group shows comparatively resistance towards the deformations, and the landforms are quite resistive type. The various parameters of the landforms i.e. slope conditions, drainage, river terraces, abondoned river channels, and landsliding etc. Were used for this study, which shows that the landforms are directly controlled by the lithology and structures of the area.

Introduction

Present envisage has been carried out in the North Eastern part of Srinagar, Garhwal lesser Himalaya between lat. 30°13' : 30°23' N and long. 78°45' : 78°55' E. The Himalaya which is perennial reserve of the water for most of the Himalayan rivers. The most imposing landscape in this area is represented by the Alaknanda river and its tributaries i.e. Badiargad, Bhadrarigad, Khankragad, Helangad, Lastergar, Devalgad, Lostugad and Switnala etc. The main stream Alaknanda possesses an anticident drainage, characterised by the presence of narrow and deep gorges carved almost at right angles to a number of parallel ridge of the lesser Himalaya.

The various landforms have been studied in the field, were mapped in detail on the base map, topographic sheet (Survey of India) No. 53 J/15 and 53J/16 and were verified through the study of landsat imageries No. 038-137 and 038-138.

For the study of slopes a new method has been followed for the slope mapping and types of slope has been discussed. The river terraces and abondoned river channels have also been studied in detail, along with the study of planar surface and lineaments of the area.

Morphogenetic Unit

The morphogenetic units have been identified on the basis of the criteria suggested by Penk (1973). In this region the climatic conditions are controlled by the altitudes, the localities at low altitude have higher temperature, low rainfall and low wind velocities, where as at the higher altitudes have a low temperature with occasional snow fall and a higher rainfall along with a very high wind velocity.

Based on the Penks classification the present area is divided into three main climatic zones.

(a) Subnival zones

The area at higher altitudes, above 2500 m to have subnival climate with some snow fall along with frozen ground and a scanty rainfall, with a very high wind velocity.

(b) Phreatic zone

Altitude between 1500-2500 m may have a phreatic climate characterised by a medium to high rainfall with occasional snow fall in winter.

(c) Subphreatic zone

The areas at low altitudes, below 1500 m have subphreatic climate, medium to low precipitation, summers are hot and winters are cold with foggy and cloudy environment.

The distribution of the precipitation in this area is controlled by orographic conditions. The monsoon arrives here from the south and SE and when comes in contact with the south facing slopes of the ridges, it is condensed and gets precipitated. The north facing slopes of the ridges come under the shadow zones and receive a very little rainfall. Since the monsoon is condensed at about 2000 m, therefore there is a little rainfall above this height and the precipitation takes place in the form of snowfall and cloud bursts.

The three distinct morphogenetic unit worked out for this area are as following :

i. Summit Surfaces

The submit surfaces are located generally above 2000 m, mainly consisting of slightly convex, more or less flat summit surfaces of the ridges. It is characterised by frost action, weathering and gravitational processes and it usually consists of weathered rock fragments, some older morains have also been noticed such as at launga village, situated along the Helaungad. The Helaungad valley is U shaped containing morainic deposits, all along its course. These surfaces constitute the water divides for various major streams in this region.

ii. Hillside Slopes

This unit ranges is altitudes from 1000 m to 2000 m consists generally of moderate to steep hill slopes with colluvial deposits and a high rain fall. The maximum rainfall is recorded between 1500 to 1800 m with occasional snowfall during the winter. Chemical weathering (Solution, Oxidation and hydration) is a dominant degradational process in this area also with some mechanical weathering (i.e. insolation and solifluction). The weathered rock fragments slide down from the steep slope and rest at an angle of repose, constituting the alluvial deposits. The alluvial deposits are found in those places where their removal is prevented by surface runoff. Therefore this unit contains pebbly as well as residual soil (Palaeosols). This unit is most widely distributed in study area.

iii. Valley Surfaces and Terraces

The main valley is widened probably due to effect of tectonically weak zones below 1000 m. The Alaknanda as compared to other areas is wider around Srinagar with appreciable meandering (Plate). The meander belt commonly accompainies depositional as well as erosional terraces located below 1000 m altitude forming the valley surfaces. This unit is characterised by gentle aggradational slopes modified on the scouring action of the surface runoff, creep and wash. It appears to possess some ground water conditions as evidenced by the presence of a number of springs. The main processes of weathering are manifested by debris avalanches on the slopes and gullies are marked through unconsolidated sediments of the depositional terraces.

Slope Conditions

Based on the field observations of the slope and measurements of their direction as well as amount, and detailed studies of topographic maps and landsat imageries, the following slope parameters were identified.

a. Slope inclination

The slope angles are measured in the field at scattered points and were plotted on the map. In the slope map thus prepared, the inclination of slope and directions were plotted on the grid pattern and the area with common direction of slopes was identified. In the southern part of the area viz. South of the Alaknanda river the slopes are generally inclined towards north and are at high angles varying from 20° to 65°, where as the south facing slopes are ranging in inclination between 20° to 50°. There are steep cliff becoming gentle away from the valley (Table 1). On the basis of inclination Strahler (1952) has divided slopes into three-

(i) Well cohesive slopes over 40° (ii) Repose slope 32° - 40°
(iii) Wash slope below 32°.

b. Slope forms: Cliff, convex, concave, gentle, straight are the main forms of slopes

The cliff type of slopes are composed of cohesive material, but some may be of repose type, observed near the river. Around Srinagar, these slopes are concave in upper part becoming straight in the middle are concave towards the base. Similar slopes have also been encountered in the tributaries of the Alaknanda viz. The Gostugad, Bhadrarigad and Badiargad etc. Repose type of slopes appear to have been formed due to the accumulation of talus material infront of the scarpment. Some convex type slopes are also observed around Chirbatiyakhal and Mayali, located just near the Summit surfaces.

c. Breaks in Slope Profiles:

The breaks of the slope have been identified on the basis of study of longitudinal stream profile. For this purpose, longitudinal profiles of the IVth and Vth order streams were prepared. The breaks of the slopes are being given in the Table 2.

TABLE 1: Representing the Slope Inclination, Slope Type and Rock Types

Localities	*Slope Inclination*			*Type of Slope*	*Direction*	*Valley Features*	*Rock types*
	Max.	*Min.*	*Mean*				
Srinagar	75°	10°	42°30'	Concave	Nfacing	Wide valley	Phyllite
Toneta	80°	20°	50°	Steep/cliff	N to NNE	U shaped	Gneiss granite
Thapla	75°	15°	45°	Concave	SW	V shaped	Gneiss
Birangaon	80°	15°	41°30'	Cliff	North	U shaped	Gneiss basic
Launga	85°	20°	52°30'	Cliff	NNE	U shaped	Gneiss basic
Mayali	85°	30°	57°30'	Cliff	South	U shaped	Gneisses
Chirpataya Khet	75°	25°	50°	Cliff	SE		Gneisses
Serra	70°	25°	47° 30'	Straight	SW	Gorge	Quartzite Limestone
Badiargad	75°	25°	50°	Cliff	ESE	Gorge	Quartzite + limestone
Silkakhal	80°	50°	65°	Cliff	E to SE	—	Quartzite
Chauras	45°	10°	27°30'	Gentle	South	Terrace	Alluvium
Kilkleswar	55°	10°	32°30'	Gentle	South	Terrace	Alluvium + gneisses
Srikot	60°	20°	40°	Concave	N to NW	Terrace	Alluvium Phyllite
Khankra	65°	35°	50°	Cliff	North	Gorge	Quartzite
Dungri Panth	60°	15°	30°30'	Concave	North	Terrace	Metabasic+ Quartzite

The above table representing the distribution of slope with their maximum minimum and mean slope in the type of slopes, valley features and Rock types.

TABLE 2: **Breaks of Slope in the Longitudinal Profile of Streams**

Number of Break	*Height of the Break in slope*	*Name of stream/s*	*Rock type*
First break	700-1000 m	Helaungad, Bhadrarigad Badiargad, Gostugad Khankragad etc.	Crystalline Quartzite, Lime-stone slàte Phyllite Metabasic
Second break	900-1400 m	Helaungad Bhadrarigad Badiargad, Devalgad etc.	Crystalline, Metabasic, limestone, slate phyllite
Third break	1200-1900 m	Helaungad, Bhadrarigad, Badiargad, Nailchamigad	
Fourth break	1500-2000 m	Helaungad, Bhadrarigad, Badiargad, Nailchamigad	
Fifth break	2000-2500 m	Helaungad, Badiargad Nailchami, Bhadrarigad etc.	

d. Slope and Geology

The slope inclinations are controlled by the character of the bad rock and its structural configuration. All along the fault zones the rectilinear steep slopes are observed.

TABLE 3: **Correlation of Slope with Geology**

Location	*Crest*	*Slope Middle*	*Foot*	*Rock type*
Srinagar	Convex	Concave	Gentle	Phyllite Alluvium
Chamdhar	Straight	Convex	Straight	Dolomite-Quartzite
Mayali	Straight	Convex	Straight	Gneiss, granites, schists
Chirpatayakhal	Straight	Concave	Straight	Gneiss, Granites schists
Badiargarh	Convex	Straight	Straight	Quartzite Dolomite basics
Dhari	Straight	Concave	Concave	Metabasic, Quartzite
Silkakhal	Concave	Convex	Gentle	Metabasic, schist, Quartzite

Drainage Characters

The drainage was analysed end. It is noticed that every morphogenetic unit contains a characteristic drainage pattern as described below:

i. Dendritic Drainage

This is observed in the upstream part of the Badiargad and Helaungad. It is characterised by the irregular branching of tributaries almost at any angle.

ii. Trellis

It is observed in part of the Helaungad and in Lastergad and is characterised by the Subparallel streams, usually aligned along the strike of the rock formations or between parallel or nearly parallel topographic features. The major streams frequently make nearly right angle bends to the ridges and their primary tributaries meet at right angle with the secondary tributaries.

iii. Barbed Pattern

The Badiargad joins the Alaknanda river at Paprasu in the form of boat hook bend and Khankragad also join the Alaknanda river in same manner forming a barbed pattern.

iv. Rectangular Pattern

The tributaries join the main stream at right angled bands, mainly controlled by structural features of the rocks.

v. Radial and Centripital Pattern

It has the streams diverging from the central elevated part. The centripetal pattern shows the drainage lines are converging into the central depression. The Rajbonga hill top having the radial drainage pattern: the Helaungad, Nailchamiga, Badiargad and Bhadrarigad etc. were originated from the different slope of ridge. The Badiargad and Helaungad both have a centripetal drainage pattern.

TABLE 4 : **Drainage Pattern**

Sector	*Drainage type*	*Dedrock and structure*	*Slope inclination*
I	Angular Obsequent Trellis Braided	Phyllite only Jointing	Moderate to Gentle
II	Trellis Dendritic	Quartzite, Dolomite Slate, Phyllite/schist Metabasic	Moderate to steep
III	Barbed Rectangular Parallel Dendritic	Quartzite, Dolomite schist, slate Metabasics	Moderate to steep
IV	Rectangular Angular Trellis Dendritic	Quartzite, Granite, Gneiss, Metabasic schist	Steep slope

vi. Braided Pattern

The main stream divided into two or more channels around alluvial is land. It is well developed in Alaknanda valley around Srinagar.

vii. Parallel Pattern

The Devalga, sweet gad, Gostugad, Khankragad joins the river in a parallel position to the Alaknanda river.

Stream Channel Characters

The configuration of the channel cross section is controlled by the characters of the bedrocks through which they have been carved. In the present work, cross section and channel patterns have been drawn from the topographic map to determine their shape and sinuousity. Two types of the profile have been drawn for this purpose— (a) channel cross section (drawn across the channel), (b) Longitudinal profile (drawn along the direction of flow of channel).

River Terraces

The Alaknanda system has a major tribuaie lastergad, Helaungad Badiaragad, Bhadrarigad, Devalgad Khankragad etc. The Alaknanda emerges from a Himalayan glacier and descends in a sinuous to meadering channel pattern. Along its course it makes deep gorges with the formation of fluvial. Fluvioglacial, and glacial terraces representing the different places of quatenary sedimentation.

a. Classification of the Terraces:

Based on their depositional environments the terraces have been classified as under:

(i) Fluvial (ii) Fluvio-Glacial and (iii) Glacial

i. Fluvial

These are well developed all along on both the flanks of valleys which may represent valley floor and flood plains that were abandoned by the channel with time. The full sequence of the river terraces is rarely preserved at all places except at Srinagar Kaliyasaur and Dungripanth.

The terraces may be paired or unpaired, consist of boulders, cobbles and pebbles of quartzite schist, granites gneisses limestones, slates, phyllite and metabasics. The fabric element exhibit a preferred orientation and are well rounded fairly spherical and are occasionally oblate, prolate and bladed. The sediments are poorly sorted. The sands of fluvial terraces are finer than those of other terraces (Rawat, 1983).

TABLE 5 : **The Elevation of Prominent Terraces of Alaknanda River Around Srinagar Garhwal**

No. of terraces	*Srinagar*	*Dhari/Kaliasaur*	*Remarks*
T_0	470	480 m	Present river bed
T_1	520	536.05	S S B Village or Kansvardhimi
T_2	574.53 m	568.90	Township Srinagar
T_3	602.28	677.50	University campus
T_4	626.45	692.20	Irrigated agricultural field (Dang village)
T_5	671.52	702.00	Dang
T_6	692.83	728.50	Located at higher localities

ii. Fluvio Glacial

These terraces are well developed along the Alaknanda and Mandakini valleys, sandwitched between the terraces of fluvial and glacial domain, such terraces are located at Toneta and Badiargarh etc. It consists of reworked sediments derived from the morains and are characterised by a heterogeneous association of rock fragments, which displays a moderate degree of sphericity and roundness, poor stratification very poor preferred orientation pattern and is devoid of cyclic sedimentation. The sandy sediments of fluvio-glacial domain in general are coarser.

iii. Glacial

In the upper courses of the Alaknanda, Mandakini and their major tributaries, there is a conspicuous development of the glacial terraces. The glacial sediments include a heterogeneous mixture of angular and subangular boulders, cobbles, pebbles and fine to very fine sands. The fabric element do not exhibit preferred orientation. The coarser clasts are mostly of granites, gneisses, schists and meta-basics. The finer sediments consists of subangular to angular grains of quartz, undecomposed felspar, mica and rock fragments, i.e. granites, gneisses, schists and metabasics. Entrapped lenses of sand exhibiting lamination and cross lamination are seen are very pooly sorted and display an erratic asymmetry in grain size toward coarseness as well as fineness. (Khan et al. 1982).

b. Development of Terraces

The Alaknanda river possess six terraces, of which four are well developed, besides the present day flood plain T_0. They have been designated T_0-T_6, where the T_6 is oldest and T_1 is the youngest . The terraces are separated by intervening scarps. These are depositional type may be paired or unpaired, showing disposition along the length of the valley. The entire sequence of the terraces is well developed around Srinagar.

Around Srinagar some of the terraces show pairing disposition, which indicates the increased rate of uplift in headward and frequent climatic change in early phases of the deposition in the Holocene time (Khan et al, 1982). The terraces T_2, T_3 and T_4 are widely developed, whereas T_5 and T_6 are found to be highly denuded due to the extensive and prolonged erosion and are only represented in the form of the small caps and lenses in the valley.

c. Evolution of the Terraces

The Quaternary period was heralded by widespread glacial activity in the Himalaya, whereas with the advent of the Holocene. The glaciers, in Himalaya began to recede to higher elevation and as a consequence large volumes of water impounded in the glaciers were released in thè major Himalayan rivers. Attempt has been made to trace out the evolution of the fluvial landforms of the Alaknanda and its tributaries during the Holocene period.

Abandoned Channels and Epigenetic Gorge

At places the Alaknanda river has abandoned their original courses to flow through newly carved out gorges, which are due to their later origin have been designated as epigentic gorges (Heim & Gansser, 1939) and the original flow is now abandoned. These gorges have been termed as fossil valleys (Witter, 1962.) whereas a river flows through an epigenetic gorge a fossil valley may be present on one of its flanks and the two are separated by a wedge of insitu rocks of variable dimensions separates the epigenetic gorges and fossil valley. In the present work these have been identified on the basis of the sediment distribution of difference in the elevation of present and previous river bed etc.

Following instances of flow through epigenetic gorges have been observed:

(i) Kaliasaur/Dhari

The river Alaknanda flows through an epigenetic gorges near Kaliasaur, at about 18 km. upstream from Srinagarh. The original course of the river was on the left bank of the epigenetic gorge, which is situated presently adjacent to the Dhari Devi temple. The same valley was separated by a hump of insituerocks. Previously this river was flowing through a very narrow gorge on the Kaliasaur side.

(ii) Pharasu

This village is situated on the upstream side of Srinagar along the Badrinath road (about 10 km.) In the past this river was flowing towards right side of the present course. After the tectonic disturbances that have been shifted towards left the difference between the abandoned valley or fossil valley is 100 metre from the present level.

(iii) Supana

The Supana village is located at about 6 km upstream of the Srinagar along the Alaknanda where as the level of the fossil valley, is 50 metre above the

Geomorphological Map Of Northern Srinagar Garhwal

present river bed. The rocks on this site are vertical quartzites with metavolcanics and phyllite. The river has shifted towards the left bank.

(iv) Srinagar—Khola

This abondoned channel is approximately 280 m above from the present level of the river bed, near the village Khola, about 1 km. south-east of Srinagar.

It is carved on the phyllitic rocks, comprises of the fluvioglacial sediment. The prominent river terrace T_6 is developed on the left bank of this abandoned channel, on the Khola village side and towards Srinagar side, the river appear to have shifted about 1 km, from in its original course in this area.

Lineament Zone

The term lineament indicates any linear structure expressed in an aerial photography or in Landsat imagery.

TABLE 6: **Correlation Between Lineament Zone and Geology**

Lineament zone	*Trends of*	*Mean altitude of unconsumed surface in m*	*Structures*	*Rock types*
Srinagar-Khola	WNW-ESE	800-1400	Parallel to NAT thrust as result of faulting	Phyllite
Silkakhal-Dhari	WNW-SE	1000-2000	Parallel to NAT & Result Of Chhantikhal fault	Quartzite Dolomite limestone Phyllite
Budhari	NNW-SSE	along nala	Paralled to Bhadrari fault	-do-
Badiargad	NNW-SSE	along nala	U to Badiar fault	-do-
Helaun	NW-SE		Parallel to Helaun fault	Crystalline gneiss, granite schist
Chirpatayakhal	N-S-to NNE-SSW	2755 to 2500	across MCt	-do-

It may be a joint fracture or fault or thrust zone. They are rarely visible on the ground (Lattman and Matzke, 1961). Prudhvi Raju and Vaidyanathan (1981) have described these fractures/fracture traces in the Easternghat area.

In the present work the most of the lineament shows preferred orientations in (i) NW-SE (ii) NE-SW (iii) NE-WSW and (iv) N-S directions. These lineaments have the relationship with regional structures of the rocks and the deformation history of the area. Prasad and Shandilya (1981) and Prasad et al., (1982), opined that the lineament with preferred orientation in N-S and NNe-SSW direction may be younger than the orogeny during which the rocks were deformed.

From the comparison of the frequency Azimuth (Rose) diagram of lineaments in this area. It appears that there is a significant parallelism between the lineament zone with the fracture traces on the rocks imperceptible reactivation along some of these fractures is inferred based on the anomalous behaviour of the meander bend in the Alaknanda and its tributaries. Hence the lineaments are thought to be surfacial expression of fractures in the underlying bed rocks propagated upward perhaps due to morphotectonic activities.

A study of lineament zones in this area as given in the Table 6 indicate that the lineaments are concordent with structural set up of the area e.g. Srinagar-Khola and Silkakhal-Dhari lineament are parallel to the NAT, showing that they might be related to the NAT. In other areas like Badiargad, Helaungad, they are parallel to be transverse fault (i.e. Badiar and Bhadrari faults), where as the lineament at Rajbonga reflects the N-S lineament, cutting across the MCT. It appears that this N-S lineament may be youngest as it has dislocated other earlier trends. The structural analysis (Shandilya 1984) also shows that the latest phase of folding has N-S trends, therefore it appears that these lineaments are structurally controlled. It also suggest that the land-forms in this area have strong structural influence.

Bhatta Rai (1981) has suggested that the location where many lineament cross each other, may be favourable locals for ore deposits. In this area also as given in the Figure they may locate near Badiargad, Launga, Shilgaon, Dahri and at Bansi may be considered for the exploration of ore deposits.

Planar Surfaces

The planar surfaces may generally be flat surfaces easily recognisable either in the field or on the aerial photographs/or landsat imagery. These surfaces indicate at erosional surfaces related to a particular landform and may show slight inclination (5° to 10°) towards the valleys. They may have a thin mantle of regolith or residual soils. The planar surfaces were recognised on the basis of study of landsat imagery for this area. They are being described in the Table 7.

TABLE 7: **Correlation of Planar Surfaces, with Geology and Slopes**

S.No	*Name of the surface*	*Altitude in m*	*Slope of the Surface*	*Rock Type*
1.	Srinagar-Chamdhar	1200-1500	0-10	Phyllite & Quartzite limestone, slate and metabasics
2.	Majriyakhal-Dhari	1600-1900	0-15	Quartzite slate metabasic dolomite
3	Regoli Kothi	2000-2400	0-12	-do-
4	Chirpalayakhal-Raj Bonga	2400-2800	0-18	Gneisses, schists, metabasic, granite

Table 7 shows that there are mainly four levels of planar surfaces i.e.(i) 1200-1500 m as located in the southern part of the area, existing over mainly the phyllite and quartzite of Chandpur group; (ii) This is located between the altitude 1600-1900 m, expressed in central part of the area, existing over the quartzite, slate limestone and metabasic of calc group; (iii) ranges from 2000-2400 m also located in the central part over the Pratapnagar quartzite and metabasic rocks; (iv) ranging from 2400-2800 m as expressed in the northern part of the area over granites, gneisses, schists and metabasics of crystalline group.

These four levels of show a very gentle inclination except that of Chirpatayakhal-Rajabonga. It is commonly recognised that the older surfaces have greater inclination and are placed on higher altitude. Therefore the Chirpatayakhal-Rajbonga represent the older erosional surface in this area whereas the surface around Srinagar may be youngest.

Landslides

The slope failure are quite frequent in this area resulting into soil creep, debris avalanches, and rock falls and slides . This failure depends on the nature of slopes, soil cover, vegetation and climate along-with the nature of rock formation-underlying the slope. They are also controlled by the masswasting created on erosional processes either on fluvial or gravity action. *Kaliasaur Landslide* is one of the most prominent landslide in this area is located at Kaliasaur (about 18 km from Srinagar). The landslide area was mapped in detail by the microgrid method. The investigation reveals that the slide is resulted owing to the presence of fracturing shearing and feather jointing caused by Chhantikhal fault in this area. This fault appears just adjacent to a meander bluff of the Alaknanda. Due to the presume of Chhantikhal fault which has crushed, sheared and fractured the rocks, during monsoon time, on the slope some new drainage has been formed on the crown part of the slide. Besides the surface run off, ground water movement is also a major cause in the Triggering of the slide in monsoon.

This landslide can be controlled partially, if the ground water movement is checked either by low pressure grouting or by construction of cut off wall/ check dam to divert the flow fo ground-water. For the slope failure gabion structures along with retaining constructions.

REFERENCES

Doornkamp and King, C.A.M., 1971. Numerical Analysis in Geomorphology. Edward Arnold. 6-15.

Heim, A. & Gansser A., 1939. Central Himalayan Geological Observations of Swiss Expedition, 1936. Mem. Soc. Helv. Sci. Nat. 73 (1), 246, Zurich 1939.

Khan A.A., Dubey, U.S. Sehgal, M. N. and Awasthi, S. C., 1982. Teraces in the Himalayan Tributaries of Ganga in U.P. Jour. Geol. Soc. Ind. 23, 392-401

King, C.A.M., 1966. Techniques in Geomorphology. Edward Aronold-London.

Penk, 1973. Attempt at Classification of Climate a Physiographic Basis. Climatic Geomorphology Mc Millan, London, 51-60.

Prudhviraju, K. N., and Vaidyanthan, R., 1981. Geomorphology of Vishakhapatanam A.P. Jour. Geol. Soc. Ind. 19(1), 26-34.

Prasad C., Shandilya, A.K. and V.K. Verma, 1981. The Lineament & Planar Surface Studies, Based on Landsat Imageries of Garhwal Himalaya. Jour. Nep. Geol. Sc. 1(2) 1981, 18-36.

Prasad, C., and Verma, V.K. , 1975. Studies in Basin Morphometry Slope Development in Western Doon Valley, Garhwal Himalaya. Him, Geol. Vol, 5, 425-434.

Shandilya, A.K. and Prasad, C., 1981. The Geomorphic Evolution of the Lineament Zone in Garhwal Himalaya. Him. Geol. Vol 12, 31-39.

Shandilya, A.K., 1984. Studies on Structures and Morpho Tectonics Between Srinagar Thrust and Alaknanda Fault, Garhwal Himalaya. Unpublished D. Phill thesis, Garhwal University.

Shandilya, A.K. and Singh S.C., A Slope Mapping in the Part of Garhwal Himalaya. Geog. Soc. of India, Vol. 48 (iii), 55 to 59.

Strahler, A.N., 1956. Quantitative Slope Analysis. Bull. Am. Geol. Soc. 571-596.

Thornbury, W.D., 1969. Principles of Geomorphology. Jhon-Willey & Sons, New York.

5

SOME GEOMORPHOLOGICAL OBSERVATIONS IN AND AROUND BHATWARI AREA, DISTRICT UTTARKASHI, GARHWAL HIMALAYA

N.P. Naithani and Dr. G.S. Rawat

The investigated area lies in Central and Lesser Himalayan part of Garhwal Himalaya. Geologically the area is constituted by two group of rocks viz. Central Crystallines represented by schist, gneisses, amphibolites and migmalites while Garhwal group of rocks composed of quartzites, epidiorites, and slates. Study area is characterised by highly rugged topography with steep escarpment and of deep gorges. Innumerable glacial lakes, glacial troughs, horn peaks, with a number of hotsprings are present in the area, which attract the people from all walks of life. In this paper an attempt is made to study the drainage pattern, lineament and slope type of this area.

Introduction

The area of investigation lies in Bhagirathi valley in and around Bhatwari in Lesser and Central Himalayan part of Garhwal Himalaya in Uttarkashi district. Geographically it lies between lattitude 30°40'-30°55' N and longitude 78° 30'- 78° 45' E along Bhagirathi river. Various geomorphological features were identified in the field and then plotted on base map. Different type of drainage patterns and lineament were identified on the topographic map and land set imagery of this area. Hotsprings were encountered during the field work and their temperature were measured by thermometer. The geo-chamical analysis of water samples was made at C.B.R.I. Roorkee for trace element. Valley side slopes were measured and mean-maximum slope angles were calculated as suggested by Strahler (1964). For channel slope study, the longitudinal profile of 8 major stream were drawn with the help of topographic map and break of scope in each stream were marked and channel scope and slope ratio were calculated as suggested by Strahler (1964).

Geological Setting

The geology of the area has been studied by several workers. Auden (1949) Gansser (1939) Saklani & Nainwal (1988) Dave & Gupta (1982), Thakur & Purohit (1980). The area constituted by mainly two different group of rocks, which are separated by Main Central Thrust which passes near Sainj i.e. 20 km from Uttarkashi along Uttarkashi- Gangotri road. The Garhwal group of rocks is represented by quartzites, epidiorites, schistose quartzites while central crystalline group is composed of schists, gneisses, amphibolites, and garneliferous mica-schists etc. The geological succession which was noticed in the field are as Table 1. The geological map with important geomorphic features shown in Fig. 1.

TABLE-1: **Lithotectonic Succession Around Bhatwari Area**

Central Crystallines	Upper Crystallines	- Garnetiferous Mica Schists - Amphibolites - Banded gneisses - Augen gneisse - Fine grained gneisses
	Middle Crystallines	- Streaky migmatites - Foliated mignatites - Augen migmatites - Mylonitii migmatites
	Lower Crystallines	- Quartizite muscovite schists - Biotite gneisses - Quartzite chlorite schist - Chlorite schist
— — — — — — — —Main — — —	Central — — —	Thrust — — — — — — — — —
Garhwal group	Garhwal group	- Schistose quartzites - Quartzites - epidicrites - mylonitic quartzites

Drainage Pattern

River Bhagirathi originate from Gangotri glacier at Gaumukj on the western slope of Chaukhamba at an altitude of 3,500 m and merges with Alaknanda in Deoprayag from where it called Ganga. In between it has carved out deep gorges at many places (Prasad & Rawat, 1982). The holy river Bhagirathi and its tributaries flow with high velocity. Most of the tributaries are fed by the glacier, therefore there is a continuous supply of water throughout the year.

In this area, ridges and summit surfaces form water divides. On the convex where water flow in all the direction from a single point exhibiting radial pattern. Such type of pattern is generally seen over summit surfaces viz. at Lingudhar, Pilangdhar and Kothi dhar. The parallel pattern of drainage is observed on valley sides. The over-all general drainage pattern of the area is of dendritic type, as most of Ist and IInd order streams join high order streams at low angles. Therefore it appear to be irregularly branching. The dendritic type pattern is seen near Saundhargad, Kamargad and Dingad. The main tributaries of Bhagirathi join it at higher angle thus trelles pattern is observed particularly in Andrigad, Nahargad, and Kumaltigad. The drainage pattern of the area illustrated in Fig. 2.

Glacial Lakes

The lake of glacial origin have been located above 2,500 m from mean sea level in the zone of ablation. Few of lakes are perennial while others are seasonal.

Origin of Lake Basin

Owing to ice movement small cirques were formed by glacier and later on their size increases due to headword erosion of melt water. These depression enlarged

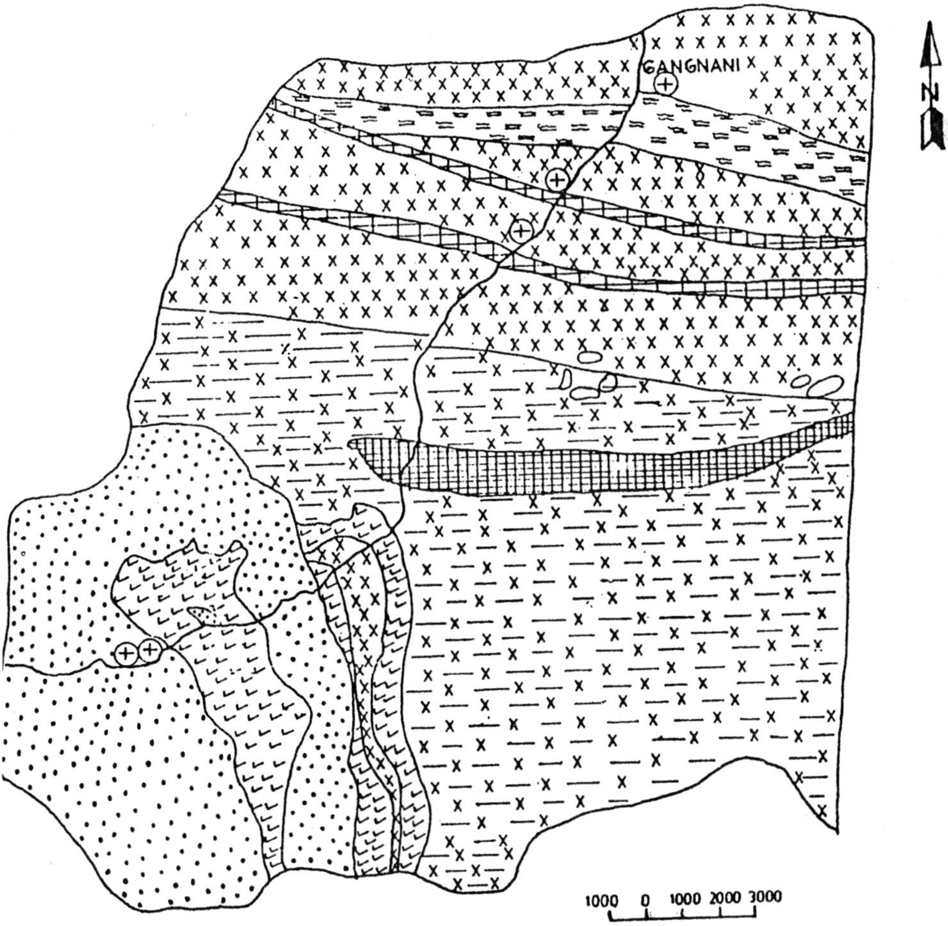

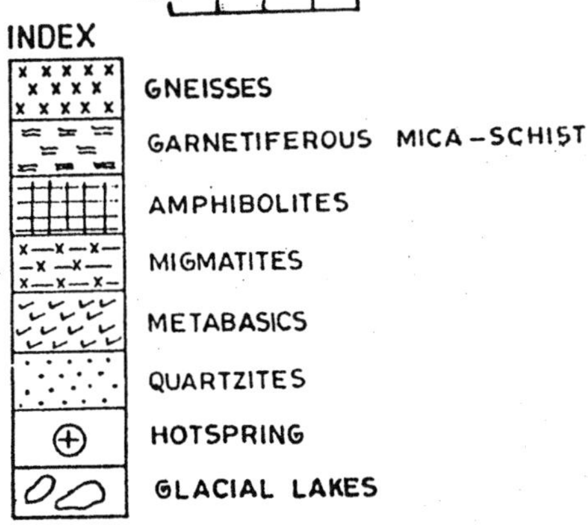

Geological Map Of The Area

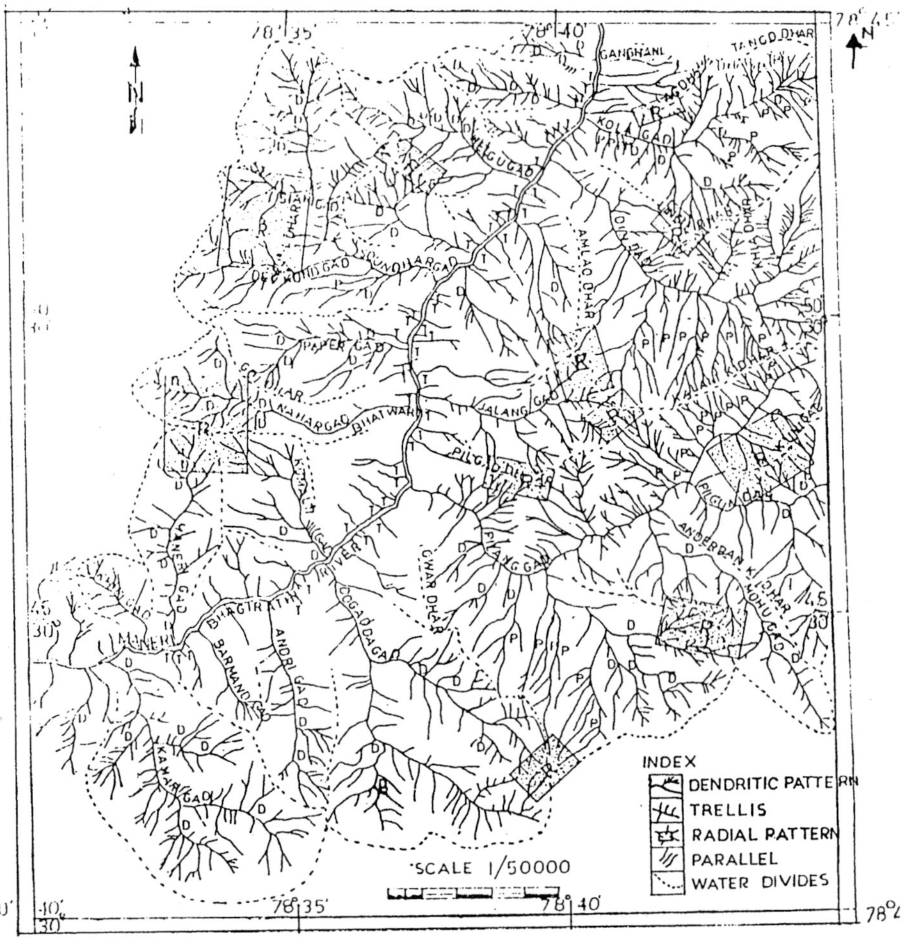

Fig 2 Drainage Map Of Bhagirathi Basin Between Maneri And Gangnani Distt— Uttarkashi

due to erosional activities of melt water and to form lake or tarn of glacial origin.

Source of Water

The area is highly precipitious. The winter season remain longer than summer. During winter and rainy season it receive greater amount of water by precipitation. In summer the ice melts and water level also increased. The water of Khariatal lake is warm may be due to the presence of hotsprings in and around the lake.

Lakes of Bhu-tap Area

The Bhu-tap area is situated about 10 kms south east of Bhukki. It is a summit surface over truncated spur. The area is occupied by thick cover of residual soil undertained by band of gneisses and amphibolites. Four lakes were observed in Bhu-tap area, one of them is perennial occupied an area of approx. 1,500 square meter. While others have areas about 375 sq. m., 450 square m and 600 square m. respectively.

Due to solar radiation and downward percolation of water these lakes become dry during summer. All these lakes are situated above 3,400 m altitude on the pasture land.

Deokund Lake

This lake is located at an altitude of 3, 100 m, covering an area of about 1,500 square m. The area around the lake is densily forested dominated by Simbru, Murinda and Soru. The bed rocks are gneisses, amphibolites and garnetiforous mica-schists.

Khariatal Lake

Khariatal lies at an altitude of 3,600 m. in the catchment area of Dingad, a tributary of Bhagirathi. The lake is circular in outline 250 m long. 230 m wide and approx. 3 to 10 m deep. The area just around the lake is densily forested. This lake is located on a vast glacial trough which is approx. 2 to 3 km long and 1 to 1.5 km wide. Innumerable glooming flowers beautify the Bugyals surrounding the lake. The place for its beauty is known as Ramnikadesh. The Jolly & Parvati peaks form the horn peaks in the area.

Lineaments

Lineaments are the linear trend of landforms reflected by the streams, ridges etc. various lineaments were identified and plotted on the map. Following major lineaments were recognised as illustrated in Fig. 3.

(i) NE-SW Lineament: Such lineament is common and is reflected by the streams, rills, ridges; Bhagirathi river from Sunagarh to Chandeti and Malla to Hinna follows this trend.

(ii) N-S Lineament: This lineament is reflected by river Bhagirathi from Gangnani to Sunagarh and again from Bhatwari to Malla. This lineament is also shown by Starigad, Amlaokidhar, and Maneri gad streams.

(iii) ENE-WSW Lineament : This lineament is reflected by Nahar gad, Jalang gad, Paper gad and Kola gad.

(iv) NW-SE Lineament: This lineament is the most prominent and has displaced

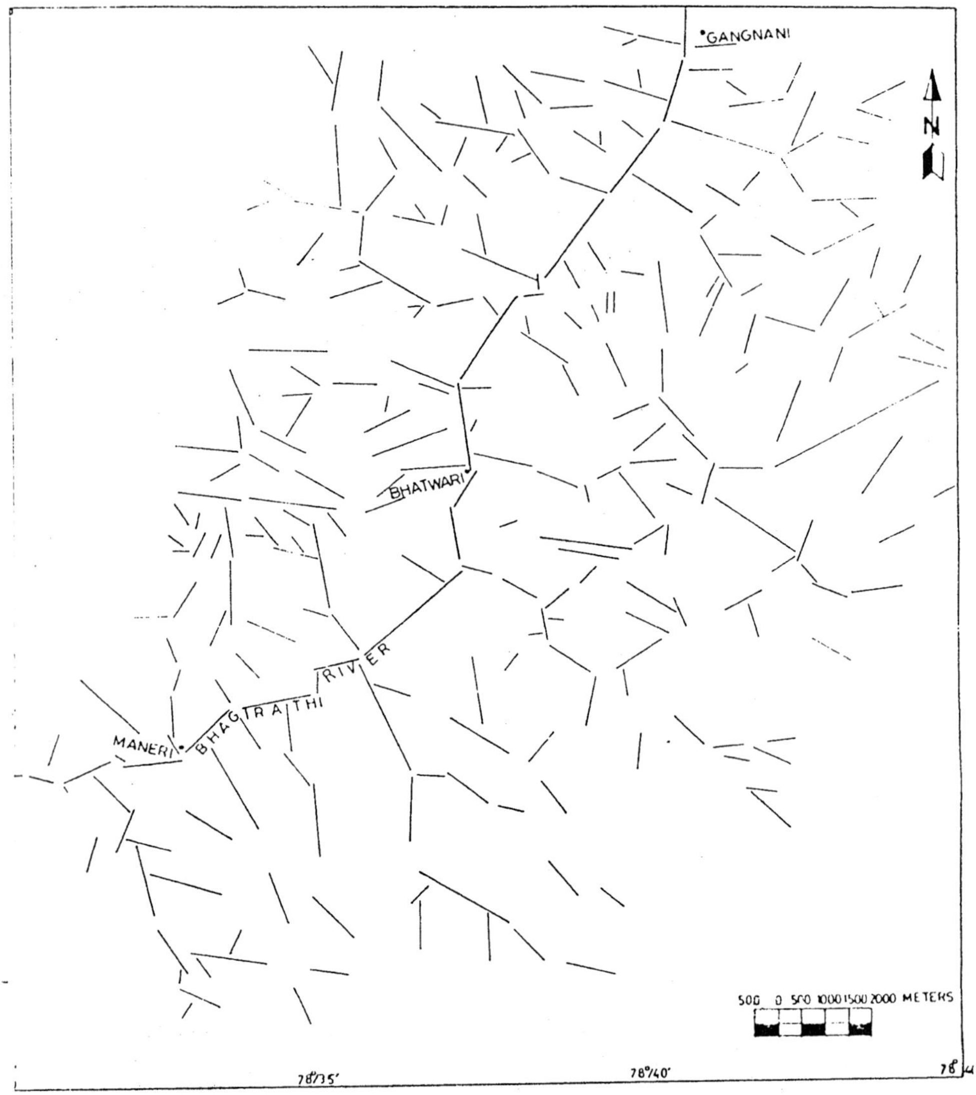

Fig 3 Lineament Map Of Uttarkashi Distt Between Maneri And Gangnani Area

TABLE 2: **Locality, Temperature and Chemical Composition of Hot Springs (ppm)**

S.No.	Locality	Temperature	Chemical Composition ppm									
			Cu	Fe	Zn	Pb	Cd	K	Mo	Mn	Mg	Na
1.	Ganganani	54°C	.54	.052	.109	—	.014	60.84	—	.117	E	E ss
2.	Sunagarh	48°C	.009	.083	—	—	.051	22.0	—	.029	E	E ss
3.	Bhukki	46°C	—	.101	.065	—	.38	30.59	.04	.021	E	E ss
4.	Maneri	36°C	.049	.109	.071	—	.093	46.61	—	.026	E	E ss

the course of Bhagirathi at many places. The tributaries such as Kumaltigad, Dogadda gad, Khiri gad, Barmand gad, Helgu gad, Din gad, Goidhar, Andarban Ki dhar etc. correspond to this lineament. Such lineament is also reflect the trend of Main Central Thrust.

NE-SW lineament appears to be the oldest as it has been deflected by other lineament. Most of the hotsprings along Bhagirathi river are located along NE-SW lineament. The N-S lineament appear to be youngest and has affected at some places the course of Bhagirathi and many of first and second order streams. The NW-SE lineament follows the trend of Main Central Thrust and is reflected by Dogaddagad and Kumalti gad. This lineament has also affected the course of Bhagirathi river at few places. As the major thrust of Himalaya trends in NW-SE direction the NE-SW lineament may represent the trend of tear faults (Prasad 1982).

Hot Springs

Hot springs have been reported at several places in Garhwal Himalaya eq. at Badrinath, Yamnotri, Gaurikund and Tapoban etc. In the present area five hotsprings have been encountered, located each at Gangnani, Sunogah, Bhukki and two at Maneri. All the hotsprings are lying along NE-SW direction on both the bank of river Bhagirathi. It seems that this is a deep fracture zone within Central crystalline rocks of Garhwal Himalaya as also suggested by Heim & Gansser (1939), Shingran & Sinha (1977). The chemical analysis of these hotsprings shows that the composition of all the hotsprings shows water is almost similar with significant quantum of K, Fe, Mn as shown in Table 2. The temperature of their water increases towards the NE-direction as noted near Maneri 36° C and Gangnani 56° C. It is inferred from the analysis chemical composition and temperature that all the hotsprings are interconnected and get water from the same source. A similar conclusion was taken by Singh & Pandey (1980) from the hotsprings of Puza valley of Laddakh area.

TABLE 3: **Valley Side Slope (Mean-Maximum Slope Angle)**

	Garhwal group	*Central Crystallines*
Crest slope	2 - 5°	5 - 8°
Mid slope	25 - 65°	20 - 75°
Foot slope	30 - 55°	25 - 56°

Slope Study

Various types of slope, slope elements and breaks in the slope have been delineated from morphometric analysis based on topographic map, landset imagery and field observations. These observations are as follows.

(a) Valley side slope: Channel slope and valley side slope provide gradient for water flow and debris transport in the fluvial system. For valley side slope mean maximum angle were calculated as suggested by Strahler as shown in Table 3. An ideal valley side slope has three parts.

(i) Crest slope: At the crest the main processes of denudation are surface wash, creep and removal by subsurface soil water. The gradient is too low in this unit. The movement is through the downward action of soil water. The angle of slope ranges 2 to 5° in Garhwal group and 5 to 8° in central crystallines.

(ii) Mid slope: This part is characterised by steep slope and rapid movement of water and debris. The steeper parts appears to be maximum slope segment of Young (cited by King 1960). Hence the removal is direct and immediate from the fall face. The slope angle ranges from 25° to 65° as observed in central crystalline and Garhwal group of rocks.

(iii) Foot slope: In this part the solution and downward movement are dominant. The steeper part of the foot slope appears to be rectilinear being cancave at some places where as in other places ending with steep and straight slope, probably due to accumulation of scree material.

TABLE 4: **Altitude and Break of Slopes**

Name of Stream	*Ist*	*IInd*	*IIIrd*	*IVth*	*Vth*	*VIth*
Manergi gad	–	3000	2800	2400	–	1800
Kumar gad	–	–	2800	–	–	1800
Dogadda gad	–	3000	2800	–	2200	1600
Pilang gad	–	3000	2800	2600	2200	1800
Kola gad	3400	3000	–	2600	2200	–
Din gad	3600	3000	2800	2400	–	–
Saundhar gad	3600	3000	2800	2400	2000	–
Helgu gad	3600	3000	2800	2400	2200	–

(b) Channel slope and slope ratio: Channel slope is defined as the ratios of vertical drop to horizontal distance measured from upper to the lower end of a single stream segment of a given order. Every slope has an inclination that is not uniform. The sudden change in inclination is called as break of slope (Leopald 1969). As depicted in Table 4 and Fig. 4. The stream profile shows six break in slope which indicate some sort of geological activity within area.

(i) The first break lies at an altitude of 3,600 m as observed in Soundhar gad, Kolagad and Dingad. The slope show steeper and straight indicate constant intensity of erosion (Penk 1979).

(ii) The second break in the slope profile is at the altitude between 3,000 to 3, 200 m as traced in Saundhar gad, Kola gad, Helgu gad, Dogadda gad, Maneri gad and Pilang gad. In case of Kola gad, and Saundhar gad, the slope is straight concave while in the case of Pilang gad and Helgu gad it is almost convex. The convexity attributed to the variable intensity of erosion of morainic materials.

(iii) The third break is located at an altitude of 2,800 m. The slope is straight in case of Kumar gad, Maneri gad, Din gad and Dogadda gad. Convex in Helgu gad and straight concave in the case of Saundhar gad. The concave slope may be due to declining intensity of erosion. The straight element suggest a parallel retreat, on accumulation of source material infront of a escarpment (Rawat 1983) as encountered near Maneri.

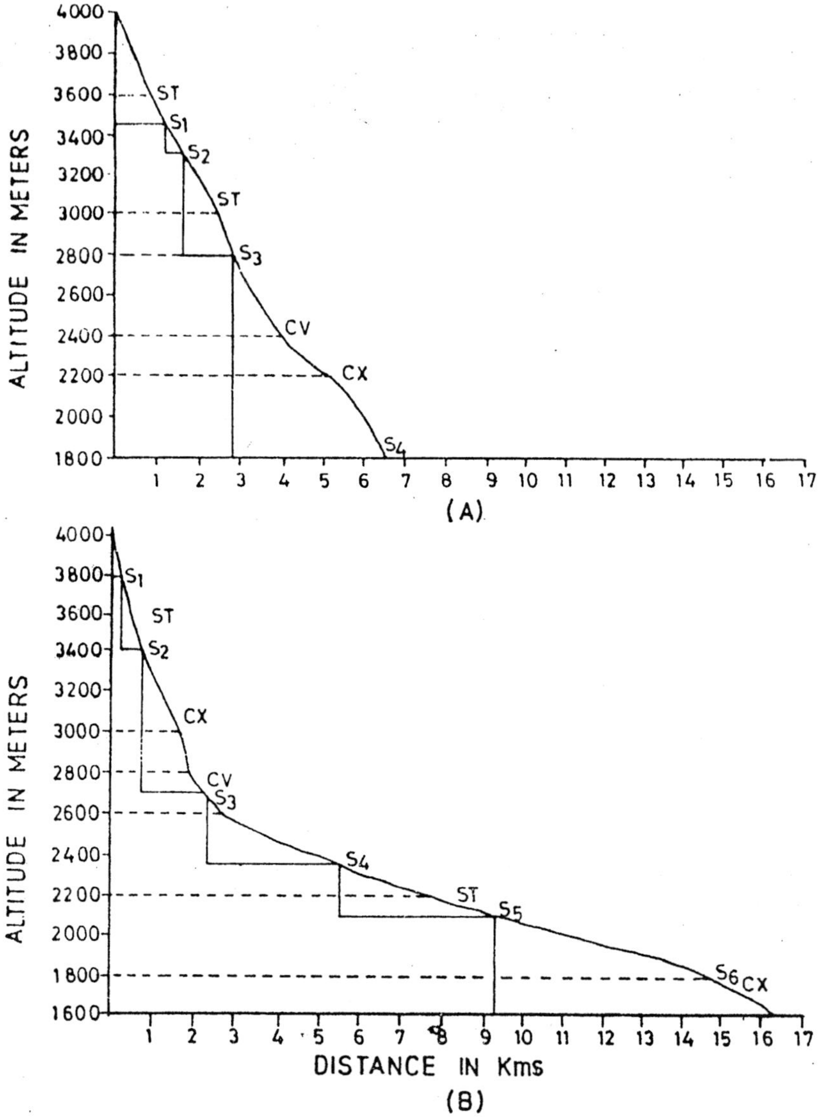

Fig 4.A Longitudinal Profile (A) Helgu-Gad. (B) Pilan-Gad.

(ST — Straight, CV — Concave, CX — Convex)
S_1–I Order Stream, S_2–II Order Stream, S_3–III Order Stream, S_4–IV Order Stream, S_5–V Order Stream, S_6–VI Order Stream,

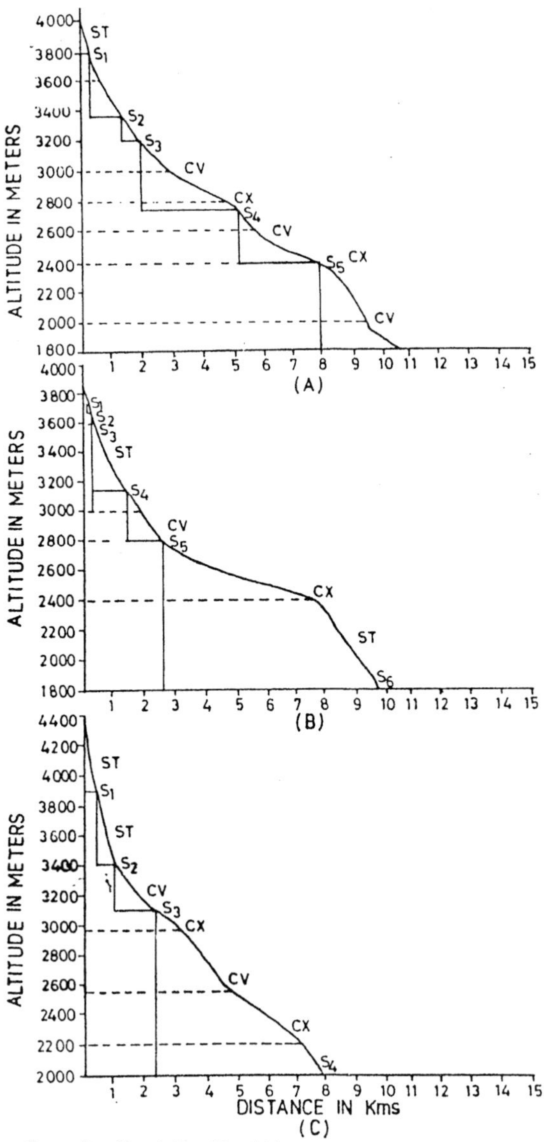

Fig 4.B Longitudinal Profile (A) Soundhar Gad. (B) Din Gad. (C) Kola Gad

(ST — Straight, CV — Concave, CX — Convex)
S_1–I Order Stream, S_2–II Order Stream, S_3–III Order Stream,
S_4–IV Order Stream, S_5–V Order Stream, S_6–VI Order Stream,

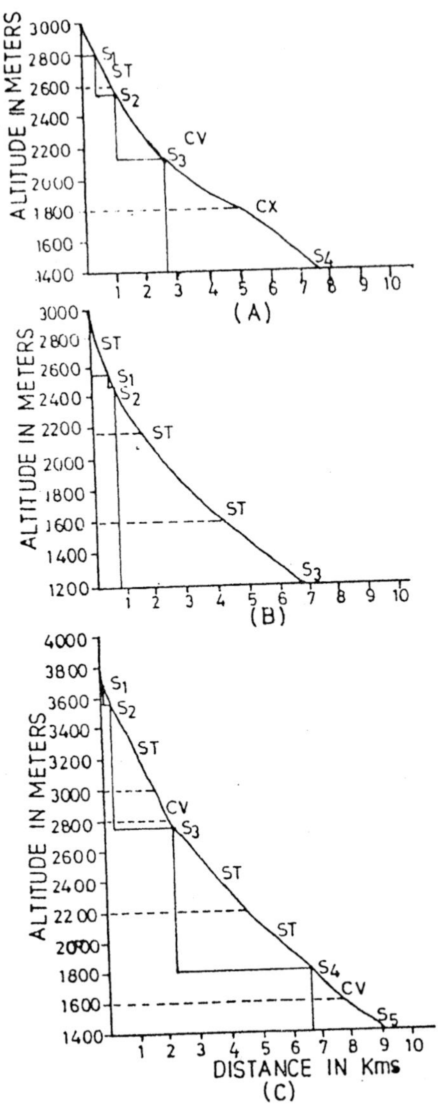

Fig 4.C Longitudinal Profile (A) Kamar Gad. (B) Maneri Gad. (C) Dogadda Gad

(ST — Straight, CV — Concave, CX — Convex)

S_1–I Order Stream, S_2–II Order Stream, S_3–III Order Stream, S_4–IV Order Stream, S_5–V Order Stream, S_6–VI Order Stream,

TABLE 5: **Channel Slopes of Various Orders of Major Streams**

S. No.	*Name of Stream*	*Stream order*					
		S_1	S_2	S_3	S_4	S_5	S_6
1.	Maneri gad	.72	.46	.22	-	-	-
2.	Kumar gad	.57	.43	.26	.15	-	-
3.	Dogadda gad	1	.93	.4	.23	.16	-
4.	Pilang gad	1	.72	.54	.10	.070	.064
5.	Kola gad	1.25	.83	.25	.20	-	-
6.	Din gad	1.3	.4	.42	.22	.14	-
7.	Saundhar gad	.67	.4	.26	.13	.13	.22
8.	Helgu gad	.49	.31	.47	.26	-	-

TABLE 6: **Mean Channel Slope and Slope Ratio of Two Unit Streams**

Stream Order	*Central Crystallines*		*Garhwal Group*	
	Mean Channel Slope	*Slope ratio*	*Mean Channel Slope*	*Slope ratio*
S_1	.94	-	.76	-
S_2	.53	.56	.60	.86
S_3	.39	.66	.36	.60
S_4	.18	.46	.19	.52
S_5	.11	.61	.16	.84
S_6	.14	1.2	-	-

(iv) The fourth break is traceable in all the streams at the altitude between 2,400 to 2, 600 m. The slope is straight in case of Dogadda gad, Maneri gad, and Kumar gad, this may be due to uniformity of climate, lithology and equal rate of erosion. The slope is concave in Saundhar gad, Din gad, Kola gad, Helgu gad and Pilanggad.

(v) The fifth break is at an altitude of 2,200 m and is found in all the streams of central crystallines. This break indicate some tectonic disturbance which may coincide with Main Central Thrust. The slope is straight in Din gad, Pilang gad, and convex in Saundhar gad, Kola gad and Helgu gad.

(vi) Sixth break is found at altitude of between 1,600 to 1,800 m in Maneri gad, Dogadda gad and Kumar gad, which is straight throughout suggesting accumulation of alluvial and scree material in lower reaches.

From the longitudinal profiles Fig. 4, it is clear that all the streams are steeper with uniform straight slope in the upper part and then approaching to straight concave. Concave in middle part and finally approaching to convexity in lower part.

Channel slope and channel slope ratio values for Ist and IInd order streams are higher in central crystallines while IIIrd, IVth, Vth and VIth order stream has higher value of channel slope and slope ratio in Garhwal group.

Conclusion

On the basis of this study following conclusion can be arrived at.

(i) The overall drainage pattern of the area is of dendritic type, the radial pattern has been developed in summit surfaces and a characteristic trellis pattern has been developed is displaced by the newly emerging quaternary surfaces.

(ii) The glacial lakes are generally present above 3,000 m. The lakes are surrounded by posture land known as Bugyals, which are coverd by blooming flowers of various colours. The Khariatal area which is located in large glacial trough can be developed to a future tourist resorts.

(iii) Delineation of lineament has revealed that NE-SW lineament is the oldest one and has being displaced by all other lineaments while N-S lineaments is the youngest and has cut all the previous lineaments. NW-SE lineament coincide with the major tectonic feature the Main Central Thrust.

(iv) Most of the Hotsprings have similar composition and lie along NE-SW lineament with temperature towards NE direction. Therefore there is a possibility to get more warm water in NE direction. The stream/warm water may be used for thermal power generation.

(v) Slope analysis indicate that upper part of valley slope and channel slope are steeper and straight while middle approaching concavity and finally convexity. The concavity and convexity in slope is due to variation in rate of erosion, presence of morainie material, scree material and structural disturbances.

REFERENCES

Auden J.B., 1949. In Directors General Report for 1939, Rec. Geol. Surv. India 78, 74-78.

Heim A. & Gansser A., 1939. "Central Himalaya-Geological Observation of the Swiss Expedition" Mem. Soc. Hal 73(1)–1–245.

Saklani P.S. & Nainwal D.C. "Metamorphitic and Migmatites of Sainj Area, Uttarkashi Garhwal Himalaya U.P. Current Trends in Geology Vol. IX Himalayan Thrusts and Associate Rocks, Saklani P.S. (Ed.) 1986, 99-122.

Gupta S. K. & Dave V.K.S., (1982). Petrology and Metamorphism of the Bhagirathi Valley Central Crystallines District Uttarkashi U.P. Hima Geol. Vol. 9 (2) 512-529.

Strahler A.N., (1964). Quantitative Geomorphology of Drainage Basins and Channel Network In V.T. Ochaw (Ed.). Hand book of Applied Hydrology 439.

Rawat G. S., (1983). Some Aspects of Quaternary Geology in Alaknanda Valley Garhwal Himalaya U.P., Unpublished thesis submitted to Garhwal University Srinagar (Garhwal).

Part II
Environmental Resources

6

RESOURCE INVENTORY AND ECONOMIC DEVELOPMENT: WITH SPECIAL REFERENCE TO RECREATION RESOURCE OF GARHWAL HIMALAYA

R.D. Gaur and D.D. Maithani

Himalaya, the sovereign mountain in the North, constitutes one of the significant resource reservoir of many natural, physical, cultural and spiritual wealth to our country. Vegetational resources are predominant and satisfy our demand upto a great extent, for various purpose at regional, national and international levels. In the recent years excessive and unplanned exploitation of well known plants has threatened the existence of life and nature of the lofty mountain ranges, consequently indicating several ecological disasters. Moreover, with the recent trends in the multifacet scope of the knowledge of the Himalaya, many of the domains need reinvestigations for suitable utilization and conservation of resources. Therefore, the developmental problems of the region require involvement of integrated resource system approach with the consideration of the ecological aspects.

Himalaya embodies several charming, potent, unique and attractive plants of ethical, historical, religious and commercial values, distributed in unique natural environments, which could be managed properly for multipurpose recreation resources to the human beings, to promote tourism by establishing nature gardens, parks and areas of 'Wilderness'.

The present communication pertains to the role of botanical studies in the developmental systems of Garhwal Himalaya, with particular reference to aesthetic and religious plants of multipurpose recreation resource.

Introduction

Garhwal Himalaya represent one of the most fascinating and characteristic vegetation, blanketing beautiful mountain ranges, valleys, gorges and ravines. It is located between the latitude 28-30°N and longitude 78-81°E, covering wide extremes of elevation range, topography, soil and slope exposures, thus supporting numerous habitats to different plant forms. There are innumerable lyrical records of vegetation and plants in proper for their various properties in our Vedic literature, which has been grossly ignored in the recent past. However, our present day knowledge on the plants of this area is due to various explorations and surveys, the important being made by Hooker (1872-1897), Strachey and Winterbottom

(1846-1849),(*see* Duthie, 1906), Osmaston (1927),Gupta (1955, 1957, 1962), Ghildyal (1956) Rao (1960), Rau (1961, 1975), Semwal and Gaur (1981) Kala and Gaur (1982) and others.

Vegetational resources of the region were thought to be unlimited and some of the resources have gained much reputation for their utilization at national and international levels, consequently exerting more economic pressure on selected taxus and at present many of the areas are under increasing human pressure for intensive resource utilization. Therefore, the development problems of the region require involvement of integrated resource system approach with the consideration of the conservation of the ecosystem on the basis of ecological, economic and social principles. In this respect recently some of the workers have enumerated plant resources and laid emphasis for necessity of biosphere reserves in the Himalayan regions (Jain & Sastry, 1978, Maheshwari 1978, 1980) . The author feels intensive resource inventory of the different sections of the Himalaya, based on ecological, economic, social and cultural aspects is necessary in consideration of future developmental programmes. Instead of looking for the routine extraction of plant resources for various economic products, many of the new facets of comparatively less or non-consumptive nature of utilization are to be explored, which is only possible through understanding of integrated resource system.

There are several aspects of developmental systems, however, the present communication pertains to the role of botanical studies in developmental systems of Garhwal Himalaya, with particular reference to recreation resources.

Recreation Resources

The recorded lyrics in the praise of beauty and grandeur of the Himalaya and its vegetation have had deep influence on our aesthetic and cultural history. Our past Vedic literature has given due credit to the vegetational wealth of the region in economic, philosophical, aesthetic and religious senses.

To apply science for amusement from such a prosperous region study of botany presents many points of interest and attraction. Though related with the living and organised nature, its application does not require tough experiments and dissections. Walking over the meadows or forests affords pleasure, a source of gratification and a spirit stirring recreation. He who has visited Himalaya amidst the blooming flowers and jungles knows well the feeling of delight connected with evanescent Nature and several of the scenic beauties indelibly impressed on his memory. That is why Lord Krishna once declares "Among immovables I am Himalaya". The Himalaya is his Vibhuti and as such are the holiest of Holies. Large number of sages and seers such as Vashishtha, Narada, Bhrigu, Karnava, Vyas and others meditated in the Himalaya and dispel their teaching to human beings. This indicated Himalaya and Holiness are inseparable in our spiritual culture. The whole of the Himalayan region regarded as above of God and Goddess, where in silent communion with Nature, man is rejuvenated and feels something super.

General types of classification of resources recreation pattern are proposed (Simmon, 1974; Usher, 1973; Singh,1977) depending on variety of biological and physical objects, but more emphasis has been laid on unique natural areas, primitive areas, historical and religious sites and high density areas. In Garhwal Himalaya, distinct pattern of natural reserves could be established on the basis of varied cultural and ecological aspects and for developing such natural resources the involvement of multidisciplinary research is essential. In this respect only some of the interesting aspects are followed hereunder.

Local Values

Local ethical systems are representing deep involvement with Nature and folk songs, religious ceremonies and proverbs are full of Nature praise. Every village, mountain range, forest is supposed to be the abode of Dieties, in simple or ceremonial ways. At the entrance of the forests some strip of white or red cloth tied to the tree or bush or sometimes heap of stones added with the plant material frequently observed, which represents forest and way side dieties. Such respect to the dieties does not indicate superstitions but a sense of gratitude to the Nature and feeling of sustained utilization while visiting to a particular forest or range. Temples or monastries are associated with the tree forms and pay equal regards to their Diety and trees. The plants associated with the temples in the lower altitude are *Ficus bengalensis, F. religiosa, Shorea robusta, Ougenia dalvergioides, Pyrus pashia* and others. The high altitude villages worship *Cedrus deodara, Morus serrata, Betula utilis Pyrus pashia,Pruhus cerasoides* and *Cupressus torulosa* with the dieties. Several flowers of local origin are regarded as sacred and offered to Dieties. The important being the species of *Anaphalis, Artemisia, Datura, Vallaris, Iris, Meconopsis, Aconitum, Plumeria, Magnolia, Saussurea, Viola* and others. Name of many villages are associated with the names of trees. Thus understanding of local values of spiritual, cultural and aesthetic is one of the important criteria of conservation and recreation in this part of country.

Monumental Tree

There are several unique trees in the region which represent ethical, historical and religious importance besides their scientific nature. Garhwal has privilege of having the oldest tree of our country. In the pilgrime centre of Joshimath one such venerable *Morus serrata* tree is present which has been studied by Rau, 1967 (see Rao, 1977). It is believed that Shri Shankaracharya, meditated under the shade of this tree during the 8th century A.D. The local people call it as'Kalpa Briksha' and worship with different aims and objectives. The author has also observed some unique forms of trees in different areas. At Gualdam in Chamoli, the oakes are of specially wide girth. One of the Deodar tree near Ban village enroute to Roopkund has about 8.5 m circumference. Resource inventory of such living monumental trees is essential for conservation as well as recreation. *Pinus roxburghii, Cedrus deodara, Cupressus torulosa, Picea smithiana, Buxus wallichiana, Betula utilis, Aesculus indica* are some of the types of the trees which attain specially large heights and girth of special interest.

Coupling of trees (two or more trees of different species jointed at the base) is regarded as 'marriage of tree' and auspicious. At Nandkeshari, Chamoli Garhwal, coupling of six trees together forms an interesting composite structure. In Raath area (Pauri Garhwal) single tree of *Pinus roxburghii* surviving in a barren slope, from far distant appears like a *Ficus* tree, it is said that when it is wounded blood ozes out (the resin is red), local people believe it an emblem of Goddess, therefore, conserve unharmed. There are several examples in the Himalaya where different trees are devoted with the name of God and specially being conserved.

Landscape

For the past landscape study was one of the important theme of geographical research but in consideration of recreation resource and conservation, it is receiving attention from many angles. Scenic aspect of a region is the result of composite image of several aspects for proper land use pattern (Kaur, 1980). Some of the workers visualised tourism value of different landscape for Garhwal Himalaya and suggested involvement of multi-disciplinary approaches (Singh, 1977). Mainly three types of the factors i.e. (i) Natural (ii) Social (iii) Historical factors have been assessed. Natural factors include scenic beauty, flora, fauna, and climate. In this respect homogenosity of the vegetation, richness of natural feature, lack of disturbing elements, panoramic views, vegetational wealth and availability of wild life in Garhwal Himalaya varies from place to place with respect to topography. Selection of the particular landscape for recreational resources thus requires intensive inventory. Simultaneously social values based on religion, festivals, fairs, architecture features together with historical and archaeological sites play significant role in landscape ecosystem.

Areas of 'Wilderness'

Pressure on vegetation with increasing population and civilization has resulted thinning of vegetation and denudation of many forest areas. However, there are some unapproachable forests and alpine meadows still representing pure natural environment and such areas may be grouped under 'Wilderness areas'. These areas are naturally protected without any deliberate management of human beings as contrast to natural reserves and gardens (Usher, 1973). These areas represent one of the ecosystem unaffected by man, distributed in the interior of Himalaya, far away from the human population that too of scarce nature. Therefore, lowest manipulation made by human beings as typified in such prinstine conditions. These are potentially useful resources and should be allowed to persist independently without any interference. These are very much useful in maintaining the gene pool of wild organisms, evolution in natural systems and allied research aspects in ecology, which can add some information of recreation resources too.

Features of Botanical Interest

Vegetational features represent richness in quality and quantity and approximately Garhwal Himalaya embodies *ca* 2500 species of flowering plants (Duthie, 1906; Osmaston, 1928, Rau, 1975) distributed over wide range of topographical and climatic situations. In this respect alpine plants bear special features of interest for their adaptation in extreme adverse environmental situations. Their comparative rarity the localities in which they grow and frequently their beautiful hue indicate distinct livelihood. Some of the extreme cold and dry habitats have cushion like plants i.e. *Thylacospermum, Juniperus, Berberis, Salix Rhododendron, Arenaria, Leontopodium, Androsace, Saussurea, Sedum* are well known for their wooly habit. There is record of plant collection from the Himalaya *(Christolea himalayaensis)* at the elevation of 6300 m by Gurdayal Singh (Rau, 1975) during Kamet expedition.

Gardens in meadows, rocks and water sheds are beautifully managed in the set pattern of gardening architect by Nature. Some of the most attractive flowers are the species of *Aconitum, Anemone, Delphinium, Geranium, Gentiana, Meconopsis, Primula, Polygonum, Caltha, Ranunculus Selenium, Saxigraga, Parnassia, Saussurea, Inula, Campanula, Lychnis, Irish, Lloydea, Fritilarea, Allium,* and *Impatiens*. Besides their miraculous medicinal values, many of these are adding to the beauty of the gardens of the world particularly in U.S.A. and Europe.

Seasonal succession of flowers, their colour, are very much characteristic in the alpine zones. The dominant colour are pink, yellow, blue, purple and white. Usually white and light and yellow flowers emit fine smell, whereas red and voilet somewhat aggreeable smell. Old literature is full of descriptions of odourferous flowers and mentioned the identity of the flowers on the basis of their smell. Change of the flower colour with different habitats in the same plant is also interesting.

Temperate and subtropical zones inhabited by different forest types (Champion and Seth, 1968) but different phytosociological association with respect to plant life and environment are so much variable, that it constitutes interesting subject of knowledge and curiosity, including several perennial reservior and wild life forms. Luminiscent type of plants in thick forests is one of the most important aspect of the aesthetic beauty. Several of the fungal members emit luminiscent light during the silent night hours. Decaying woods, some members of grasses, *Oenothera, Arum,* lilies emit luminiscent light. Instinct type of appeal of such type of twinkling light was appreciated by famous Sanskrit Kalidasa in Kumarsambhava. Such curiosity of nature has been ignored during these days.

Western Himalaya including the mountain ranges of Garhwal embodies many endemic plant species, (Puri, 1960). The rate of endemism and speciation is comparatively higher in the Himalayan region. Some of the endemic species are *Clarkella, Hemiphragma Picroorhiza, Falconaria, Nardostachys, Platystemma, Meconopsis Roylea, Vicatia,* and others. The phytogeographical affinities of the plants are very interesting. The tropical flora growing in the warmer parts of the mountains shows affinities with Indo-Malayan elements, the temperate flora represent Sino-Japanese elements, some steeppy elements are of middle asiatic origin and have their origin in low lands of Turkistan, Pamir and Afganistan. Tropical African-Madagaskar, Mediterranean, Australian, elements are also quite abundant (Mani, 1974). Recent explorations and surveys resulted several new plant species in the Flora of India. Some of the important species include *Allium, Berberis, Bidens, Begonia, Bulvophyllum, Corydalis, Circeaster, Cypripedium, Cyathea, Oenothera, Oberonia, Polygonum, Gentiana Utricularia* and others. Recent reports on *Tetracentron sinense, Hydrobryum griffithii, Mitrastemon yamnotri* (Rafflesiaceae), *Circeaster agrastris, Cypripidium* and others have been significantly interpreted on the phytogeographical aspects (Rau, 1978, Maheshwari, 1980). Several species of *Utricularia, Drosera peltata, Pinguicula alpine* are some of the interesting insectivorus plants. *Boschniakia himalaica, Aeginitia indica, Arceuthobium minutissimum. Orobanche* spp. are some parasites and *Monotropa uniflora* and several orchids are of saprophytic nature. Epiphyte plants of specific interest are *Sedum, Peperomia,* members of Loranthaceae, Orchidaceae, Lycopodiaceae, Ferns, Lichens and others.

Conclusions

Himalayan vegetation is under great stress and only some pockets represent original nature of flora (Gaur, 1982), whatever remained require proper attention for conservation.

Consideration of the aesthetic, religious and spiritual philosophy towards the mighty nature was greatly associated with the Himalaya, in the past. There is great need for going into details of Vedic philosophy for proper conservation, and continuance of old philosophies.

Recently tourism activity has been greatly increased but such developments need thorough analysis, and it is felt that tourism must be centered with pilgrimages. Various prominant places enroute to pilgrimage shrines i.e. Badrinath, Kedarnath, Gangotri, Yamunotri, Hemkund where the vegetation has been disturbed, raising of gardens of wild Himalayan plants would greatly satisfy the need of recreation resources as well as other economic aspects. Understanding of proper season of germination and development of a particular species in specific environment is essential for raising the botanical gardens.

Natural reserves and National parks must be established for the conservation with the understanding of multidisciplinary aspects. Areas of 'Wilderness', which are still away from the human activities be demarcated and for the continuance of 'Wilderness' special measures are required to remain undisturbed from the human beings.

Developmental problems of the region require intensive resource inventory and their application need ecological understanding.

The conventional form of development and extraction of the natural material must be avoided. Some new methods of nonconsumptive type of utilization are to be explored. Consumptive types of utilization should be based on farming systems. In this respect production of timber and medicine are mentionable as due to these products natural vegetation is greatly denudated, therefore, these products must be raised from farming and cultivation in suitable lands.

Recreation resources are the best form of plant utilization with respect to conservation, for which detail inventory of the spiritual, cultural, economic and ecological aspects are required and rare or endangered plants need special attention.

REFERENCES

Atkinson, E.T., 1882. *The Himalayan Gazetteer* Vol. I part III, (Reprinted Bishan Singh Mahendrapal Singh Dehradun).

Champion, H.G. & Seth, S.K., 1968. *A Revised Survey of the Forest Types of India*. Delhi.

Duthie, J.F., 1906. *Catalogue of Plants of Kumaon and Adjacent Portion of Garhwal and Tibet Based on the Collections of Strachey and Winterbottom During the Years* 1846-1849. London.

Gaur, R.D., 1882. Dynamics of Vegetation of Garhwal Himalaya. In Vegetational Wealth of Himalaya (ed. G.S. Paliwal), pp. 12-25. Delhi.

Ghildyal, B.N., 1957. A botanical trip to Valley of Flowers. *J. Bomb. Nat. Hist. Soc.* 54: 365-386.

Maithani, D.D., 1988. Possibilities of Tourism Development in S.W. Tehri-Garhwal, Gang Geographical Journal Vol. II.

Gupta, R.K., 1955. Botanical Exploration in Bhillangana Valley of Earstwhile Tehri Garhwal state. *J. Bomb. Nat. Hist. Soc. 53*: 581-594.

Gupta, R.K., 1957. Botanical Exploration in Earstwhile Tehri Garhwal State II. *J. Bomb. Nat. Hist. Soc. 54*: 878-886.

Gupta, R.K., 1962. Botanical Exploration in Earstwhile Tehri Garhwal State III. *J. Bomb. Nat. Hist. Soc. 59*: 486-512.

Hooker, J.D., 1872-1897. *Flora of British India*. Oxford.

Jain, S.K. & Sastry, A.K., 1978. Plant Resources of the Himalaya. In *National Seminar on Resource Development in Himalayan Region.*, pp. 98-107. DST New Delhi.

Kala, S.P. & Gaur, R.D., 1982. A Contribution to the Flora of Chamoli Garhwal in U.P. In Vegetational Wealth of Himalaya.(ed. G.S. Paliwal), pp. 352-419. Delhi.

Kaur, Jagdish, 1980. Tourism-Recreation Resource Analysis: fixing a criteria. *JOHSARD* 4: 101-107.

Maheshwari, J.K., 1978. Biotype Sanctuaries as an Aid to the Development Himalayan Resources. In *National Seminar on Resource Development in Himalayan Region*, pp. 146-152. DST New Delhi.

Maheshwari, J.K., 1981. Plant Resource of Himalaya and their Conservation. *JOHSARD* 4: 3-11.

Mani, M.S., 1974. The Flora. In *Ecology and Biography of India*. Ed. M.N. Mani, pp. 159-177. Hague.

Osmaston, A.E., 1927. *A Forest Flora for Kumaon*. Allahabad.

Puri, G.S., 1980. *Indian Forest Ecology*. Vol. I, II. New Delhi.

Rao, A.S., 1977. Old Trees of North-West Himalayas. *Indian Farming 26*: 9-10.

Rao, A.S., 1978. Inventory of West Himalayan Flora. In *National Seminar on Resource Development in Himalayan Region*. pp. 133-145. DST New Delhi.

7

THE MINERAL RESOURCES OF WESTERN U.P. HIMALAYA: SOME CORRECTIVE MEASURES AGAINST THEIR DEPLETION

U. S. Rawat

The Western U.P. Himalaya is a vast and dissected mountainous terrain and includes a variety of mineral resources. Depletion of these resources due to ill planned schemes can deprive the people of this region of most of its wealth in the name of preserving a small quantum. The present paper gives a wide range of distribution of various type of minerals in the region. Indiscriminate exploration of the minerals have culminated in the deterioration of the hills for which remedial measures have been suggested.

Introduction

The western U. P. Himalaya is a mountainous region which constitutes the two administrative divisions of Kumaon and Garhwal. This mountainous terrain is one of the most facinating segment of the Himalayan arc and in many ways is quite unique in its geographical and geological setting. The river Tons, separates it from the Himachal Pradesh in the west, Kali in the east, Siwalik-Bhabur foot hills in the south and Indo-Tibetan-Nepal border in the north (Fig.1). This part bears mineral resources of various types which are utilized indiscriminately and have caused an imbalance in the environment.

Physiographic Features

Physiographically this part of Himalaya is highly elevated characterised by the river Yamuna, Ganga, Ramganga, Kali with their large number of tributaries, prominent peaks, ridges, scarps, deep gorges and mountain slopes. The river valleys are characterised by a large number of paired and unpaired terraces. Land slides. water falls springs, talus cones, scree and piedmont deposits are common features.

Transversely the Kumaon and Garhwal Himalaya can be divided into four mountain belts or zones. These zones commences from south to onwards as : outer Himalaya, lesser Himalaya, central Himalaya and the Tethyan Himalaya. The stratigraphy, correlation, structure, petrology and evolutionary history have been dealt with various workers in the past decades (Pilgrim and West, 1928, Burrard et al 1934; Auden, 1937; Heim and Gansser, 1939; Gansser, 1964; Valdiya, 1980; Pande and Saxena, 1968 etc.).

Outer Himalaya

The outer Himalayan zone consist of terrigenous sediments (Molasse) of

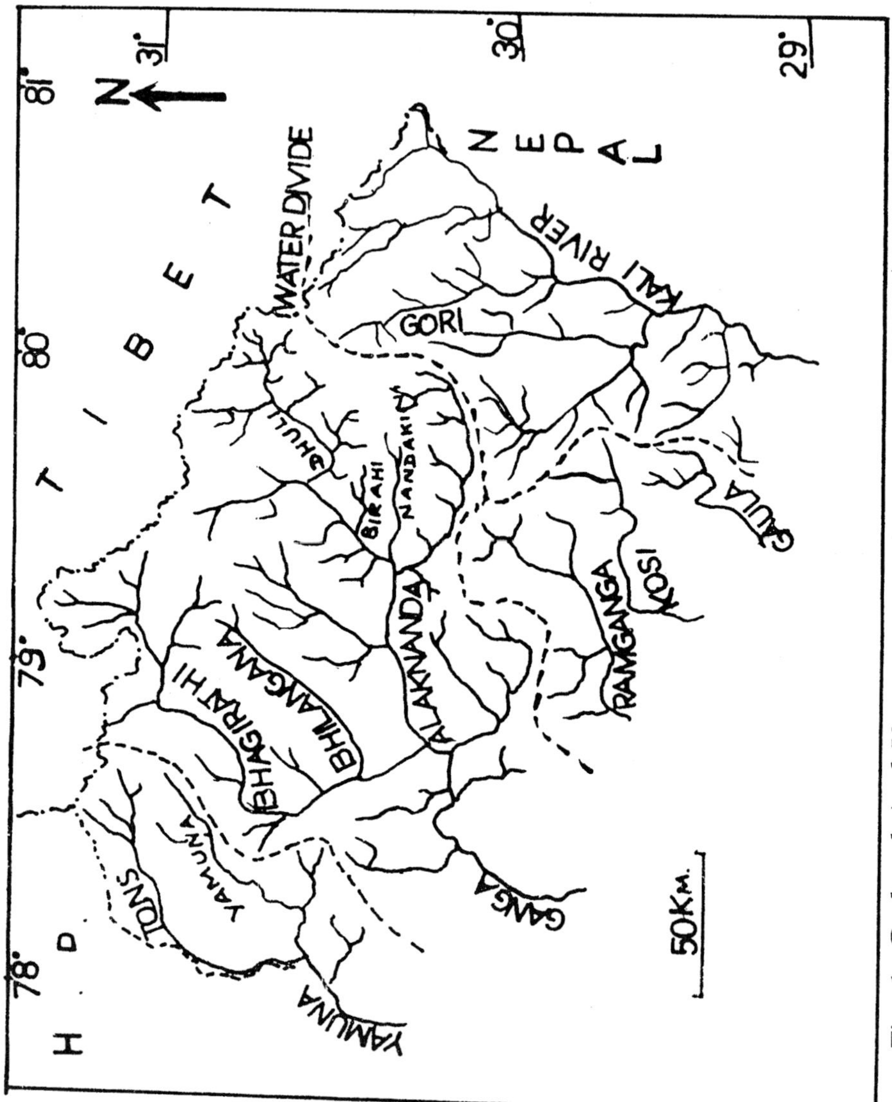

Fig. 1 Garhwal And Kumaun Himalaya, U.P.

(U.S. Rawat,Deptt. of Geology, Govt P.G. College, Rishikesh)

fluvialite environment. These sediments comprised of sand stones, conglomerates, shales and mud stones. These form the Siwalik ranges. The main boundary fault (M.B.T.) separates these from the lesser Himalaya which spreads parallel to the north of the Siwaliks.

Lesser Himalaya

This zone consists of para-autochthonous metasediments of arenaceous, argillaceous and calcareous compositions (Jaunsar, Tejam, Blaini-Krol, Tal-Subathu formations) and forms the base of the central crystallines. This zone is characterised by a number of klippen or nappes of the crystalline metamorphites which have their roots in the central crystalline unit of higher Himalaya.

Central Himalaya

This belt forms the eminent chain of Himalayan Mountains built up of highly metamorphosed rocks which are delimited by the north dipping main central thrust (M.C.T.) from the lesser Himalayan rocks. The crystallines are devoid of fossils and till today believed to be a product of remobilised basement.

Tethyan Himalaya

The tethyan Himalaya follows to the north of the central crystallines and begins with horizontally bedded & fossiliferous sediments. These sediments ranges in age from late precambrian to lower Eocene.

Climate Vegetation and Soil

The valley floors and slopes experience sub-tropical climate in general and the higher altitudes, temperate climate in particular. The higher altitudes experience snow fall commences at the end of November and lasts upto the end of April. May and June are generally hot and the end of June begins with monsoonic rains which continue upto September. October is quite pleasant and charming and the area blooms with a variety of flowers and faunas.

The soil in the region is the combined effect of Mechanical, chemical and biotic agents. The soil distribution is uneven mainly controlled by the source rocks, hill slopes, vegetational cover and other sub-aerial geological agents.

Drainage of the Region

The region is characterised by dendritic pattern and partly by trellis pattern. The area is drained mainly by a number of rivers the prominent are: Ganga, Yamuna, Ramganga and Kali. Most of the rivers and their tributaries are fed by glaciers and sprints at their sources. A number of lakes like Nainital, Naukuchiatal, Bhimtal, Saltal, Khurpatal; in Kumaon and Gonatal (lake), Roopkund and Taptkund in Garhwal are the characteristic features.

Geology

This part of Himalaya has been investigated by a number of workers. Audens classic work (1934, 1937 and 1949) has formed the base for later geologists. Heim and Gansser (1939) made regional studies. Valdiya (1975, 1980) has carried out a regional study and divided it into four major lithotectonic units. The litho-tectonic units of the area are given in the Table 1.

TABLE 1: **Lithotectonic Units Of Kumaon Lesser Himalaya**

(After Valdiya, 1975, 1980)

Outer Lesser Himalaya		*Inner Lesser Himalaya*
		Vaikrita Group (Early Precambrian)
		..........Vaikrita (Main Central Thrust).....
Almora Group	(Gumalikhet formation)	
	(Champawat Granodiorite)	Munsiari Formation
	(Saryu Formation)	
	............Almora Thrust..............................	Munsiari Thrust........
Ramgarh Group	(Debguru Porphyroid	
		Barkot-Bhatwari Units
	(Nathuakhan Formation	
	...........Ramgarh Thrust............................	Barkot-Bhatwari Thrust.......
	(Subathu Formation (Lr Eocene)	
Sirmur Group	(Singtali Formation (Palaeocene)	
	(Tal Formation (Permian)	
Mussoorie Group	(Krol Formation	
	(Blaini Formation	
	(Nagthal Formation	
Jaunsar Group	(Chandpur Formation	Berinag Formation
	(Mandhali Formation	
	................Krol Thrust...............................	Berinag Thrust.............
	Subathu (Lr. Eocene)	
		Tejam Group (Mandhali Formation
		(Deoban Formation
	(Rautgara Formation	Demtha Group (Rautgara Formation
Damtha Group	(Chakrata Formation	(Chakrata Formation

Mineral Resources

This region, representing four major belt of Himalayan arc, has been investigated in the past decades by a number of geologists. The works have shown that Mineralisation is controlled by number of magnatic and tectonic features during the evolution of Himalaya. The important mineral resources of the region are traceable in all the belts or zones. These resources are:

(a) Water Resource

Water is the basis of life and the related schemes have to be planned, keeping in view, optimum benefit to the affected populace. Many areas in the Garhwal and Kumaon are endowed with natural grandeur but can not be developed because of lack of drinking water facilities. A number of river valley projects are under investigation (Saklani, 1980) in the Garhwal and Kumaon Himalaya.

A large number of hot springs occur in the central Himalaya near Jamnotri in Uttarkashi and Badrinath and Tapovan in Chamoli Districts (Gupta and Singh, 1977). The hot springs seem to have a good geothermal potential and work for the exploration and use of the geothermal energy resources.

(b) Copper

It is an important metal extracted from its important ores like—chalcopyrite, cuprite and chalcocite etc. The copper ores, in the Kumaon and Garhwal hills, have been mined in the early period of the 20th Century before independence of India. This is evidenced by findings of old workings which were abundened by a number of factors. The sporadic occurrences have been reported from Deoban, Chail and Berinag formation.

The lime stone & dolomites of Lameri & Gauanagarh formation (Mehdi et al, 1972) in Dhanpur-Dobri area of Chamoli district contains disseminated deposits of Copper ores. The dolomite belt streching between Kaproli and Manikhath near Dagarpatti in Tehri Garhwal also contains Copper mineral deposits. The Berinag formation (Valdiya, 1968) of Askot area in Kumaon region also contains poly minerallic lode of chalcopyrite, chalcocite, cuprite besides arsenophyrite stibnite, sphalerite and galena.

Chail unit in Ramgarh and Nagnath Pokhri area in Chamoli district is characterised by poly metallic mineralized zones.

(c) Iron Ores

Disseminated seggregation and occurrances of pyrite, magnetite, haematite, limonite are found at a number of localities. Magnetite and pyrite are found associated to basic and metabasic igneous rocks. The metavolcanics of Chamoli, Bhowali and Uttarkashi district contains voluminous deposits of iron ores.

(d) Galena

Galena, an important ore of lead, is found sporadically as traces within quartzite and phyllites of Garhwal group of rocks in Alaknanda, Mandakini and Bhagirathi valley. The sporadic occurrences are also recorded in Almora and Pithoragarh districts.

(e) Uranium

Uranium, the most important and strategic mineral has been traced along the line of old copper workings in the fault and shear zones in the Chamethi-Pokhri-Tungi belt in the Chamoli District. The Uranium mineral as traces has been found (Sharma, 1974) in the Mussoorie phosphorite and black shales of the lower Tal formation.

(f) Limestone and Dolomite

The large limestone and dolomite deposits are known from the Gangolihat-Sor-Thalkedar (Mandhali) and Krol Formation of Garhwal and Kumaon hills. The best quality limestones and dolomites occur at Pipalkoti & Rudraprayag in Alaknanda Valley. Cement grade lime stone deposits have been reported from the section between Jiyaljibi and Bulwakot in the Kali river, Kumaon Himalaya by Srivastava and Redcliffe (1979). The cream coloured, hard, massive, compact and thin bedded high grade limestone are found along the Mussoorie hills. Bajamara formation of the Deoban group contain high grade limestones. The total reserves estimated (Jangpangi, et al, 1979) to be about 250 million tons. In Kumaon Himalaya medium to high grade limestone is found in Pithoragarh and Nainital districts.

(g) Phosphorite

The lower permian Jogira member of the Tal formation in Mussoorie hills, north-east of Dehradun, contains workable deposits of phosphorite deposits. The important deposits are located at Durmala, Maldeota, Chamsari, Masrana, Chaunpa-Kumali and Pari-Tiba. Across the Ganga in the Lansdowne hills, the carbonaceous shale of Tal contains phosphatic nodules. Phosphatic nodules also have been found in Kuling and Laptal-spiti shale near Neti area of Garhwal Himalaya (Tiwari, 1972). The Dhari member (Valdiya, 1972) of the Gangolihat dolomites also contains lenticular deposits of Phosphorites.

(h) Talc

Talc and its impure variety known as steatite occurs as veins, lentils and beds in carbonate rocks of Garhwal group and in Berinag formation of Kumaon Himalaya. Talc deposits of workable nature occur at Pipalkoti, Belakuchi, Heuna in the Alaknanda valley and at Simli in the Pindar river in Chamoli district. Talc deposits of workable nature are found associated within carbonate rocks of Kumaon region in Pungar, Lahor and Gori Valley.

(i) Magnesite

Magnesite is a valuable mineral utilized extensively in industrial areas. High grade magnesite deposits have been reported from Pinder valley located at Bagoli by the geologists of W.I.H.G., Dehradun. Isolated patches of high grade quality are found in the carbonate rocks exposed around Pipalkoti (Gaur et al, 1977). The Magnesite deposits of the best quality occur to the east of Ramganga, between Kali and Ramgaga and between Ramganga and Saryu river in Kumaon region. The important localities in Pithoragarh area are Bisabajer, Chandak and Dharigaon.

(j) Building Stones and Materials

Rocks of good quality can be successfully used as a dressed block or slab in different shapes and sizes in the construction of buildings, dams, retaining walls as building stones. The Sor (–Mandhali), Chandpur and Gumalikhet formation (Almora group) are the main source for buildings stones. Lose and unconsolidated sediments such as sand and gravel are used in the construction of buildings and engineering structures. The river valley are the main source. Sand and sandy type of sand stones can extensively be utilized as raw material in the ceramic or glass industries.

(k) Soil

Soil is unconsolidated material produced due to weathering agencies from the rocks of different lithological characters which, is modified subsequently by a variety of processes. Soil is the main natural resource of this mountainous region and supports indirectly the agricultural products. Mere use of improved variety of seeds, suitable for plants with fertilizers, will not really help to secure an increase yield of crops in the region. The quality and the type of soil depends mainly on the source rocks, slopes, climatic conditions, processes responsible for the soil formation etc. Proper assessment of geological aspects, would avoid wrong inputs of seeds, fertilizers etc. and make the schemes realistic and consistent with actual needs of the regions.

Resource Development and its Impact on Environment

Nature has bestowed the Garhwal and Kumaon region with rich resources. Limestones, dolomites and Phosphorite form a major bulk which are extensively and improperly quarried and mined in the lesser Himalayan ranges particularly from the Mussoorie-Nainital hill ranges. Most of the limestone and dolomite quarried are privately owned and it is in these mines that unscientific exploration and management is taking place. The stability of the hills in the area is endangered. Land slides, rock falls, solifluctions, erosion and silting of river valleys are common feature observed in the region. In the past two decades unscientific exploitation of the resources have created natural hazards. We must used the natural resources but with restraint and care and act as steward of nature to serve the best interests both present and future generations.

Corrective Measures Suggested

Improper utilization of natural resources of the region may adversely affect the ecosystem of this region and may do untold harm to the nation . For the proper utilization of natural resources the author suggests some corrective measures.

1. An assessment of natural resources of this part of Himalaya is a necessary prerequisite for the prosperity of the hill people and for the balance of maintenance of the ecosystem of Himalaya. These would call for indepth study of rocks, soil, minerals and water resources.

2. Mine lease must not be given in areas which are susceptible and geologically weak. They must be avoided in case of very steep slopes or in areas which are devoid of vegetational cover.

3. The fertility of soil must be studied before plantation and the choice of species to be planted in the area will be governed by the site, altitudes and climate, hence these factors to be studied in details.

4. Blasting if necessary during constructive work, must be used only in urgent cases but with minimum natural hazards.

5. Conservation measures for water shed management should also be brought in practice.

6. Large scale afforestation and fruit gardens should be encouraged in and around the localities where mining and quarrying is carried out.

REFERENCES

Auden, J.B., 1934. The geology of Krol belt. Rec. Geol. Surv. Ind. 67: 357-454.

Auden, J.B., 1937. The structure of the Himalaya in Garhwal Rec. Geol. Surve. Ind., 71: 407-433.

Auden, J.B., 1949. In gen. directors general reports for 1939. Rec. Geol. Surv. Ind, 78 : 74. 78:

Burrard, S.B. and Hayden, H. H., 1934. A sketch of the geography and geology of the Himalaya mountain and Tibet. Manager of publication, Govt. of India,Delhi.

Gansser, A., 1964. Geology of the Himalaya. Interscience Publishers, London.

Gaur, G.C.S., Dave, U.K.S and Mithal, R.S., 1977. Magnesite deposits of the calc zone of Chamoli, Garhwal Himalaya, U.P., India. Him.Geol., 2: 256-292.

Gupta, M.L. and Singh, S.B., 1977. The Himalayan Orogen and its geothermal resources. Him. Geol., 7: 326-346.

Heim, A. and Gansser, A., 1939. Central Himalaya. Geological observation of the Swiss expedi in 1936. Mein. Soc. Helv. Sci. Nat., 73:1-245.

Jangpangi, B.S., Dhaundiyal, J.N. and Verma, R.N., 1979. A note on the occurrence of high grade limestone in Deoban-Bajamara area, Dehradun District, U.P., Indian Minerals, 33 (4) : 1-4.

Mehdi, H.S., Kumar, G. and Prakash , G., 1972. Tectonic evolution of eastern Kumaon Himalaya : a new approach. Him. Geol., 2: 481-501.

Pande, I.C. and Saxena, M.N., 1968. Birth and Development of Himalaya. Pub. Cent. Adv. Stud. Geol. P.U., 4:1-19.

Pilgrim, G.E. and West, W.D. 1928. The structure and correlation of the Simla rocks. Men. Geol. Surv. Ind. 53 : 1-140.

Saklani, P.S., 1980. Natural Resources of Central Himalayas—A Plea against their misuse. Himalaya Man and Nature, Vol. 4 (4) : 3-8.

Sharma, K.K., 1974. Study of radioactive phosphorites of Mussoorie, Dehardun District, U.P., Him. Geol. 4:216-228.

Srivastava, R.N. and Redcliffe, R.P., 1979. The occurrence of cement grade limestone in Jiyaljibi Bulwakot areas, Pithoragarh District, U.P. Indian Minerals, 33 (4) : 48-50.

Tiwari, D.N. 1972. Nesquehonite—a possible precursor in the origin of Himalayan Magnesite deposits. Him. Geol., 2 : 94-102.

Valdiya, K. S., 1968. Origin of the Megnesite deposits of Southern Pithoragarh, Kumaon Himalaya. Eco. Geol., 63: 924-934.

Valdiya, K.S., 1972. Origin of Phosphorite of the late precambrian Gangolihat dolomite of Pithoragarh, Kumaon Himalaya, India. Sedimentology, 19:115-128.

Valdiya, K.S., 1950. Lithology and age of Tal formation in Garhwal and implication on stratigraphic scheme of Krol belt in Kumaon Himalaya. Jour. Geol. Soc. India, 16: 119-134.

Valdiya, K.S. (Ed.) 1980. Geology of Kumaon Lesser Himalaya, The Himachal Times Press, Dehradun.

8

THE CRUMBLING HIMALAYA (ENVIRONMENTAL DEGRADATION AND SOME CONSERVATIONAL MEASURES)

M. S. Anantharaman

The Himalayan saga is man's quest for knowledge, needs, adventure and spiritualism. It has been a source of inspiration and veneration for million people. "Man through the centuries, has always chosen places which seemed special and declared them holy and sacred." There lived the Gods....today, more than ever, we have a need of unexplored wilderness" writes Reinhold Messner.

True, the snowy summits have drawn for centuries, explorers, naturalists, geoscientists, environmentalists, poets, pilgrims, sages to their bosoms. For some they are a challenge and for others a pilgrimage.

The Himalayan range stretch from Pakistan to Arunachal Pradesh India with a length of 2500 kms and width of 400 kms and a astonishing mean elevation of 6000 metres at their central axial range, the Himalaya—the greatest physical features, must be one of the most spectacular and dramatic areas on the earth. Not only does it embrace the highest mountains in the world, but also mighty glaciers and swift flowing ferocious rivers, huge foot hills and fertile valleys, extensive deserts and dense forests.

And it is not only the topography that is so impressive—there is a considerable variety of mountain people with different culture, religions, languages, flora, fauna, landforms, geological structures and rocks. It is the feeling of awe and excitement that has attracted everyone. Water, soil, forests and tourism in the most part of the Himalaya has become more and more severely depleted with the passage of time, (Hillary, 1990).

WATER

The liquid mirrors of Himalaya—its rivers and lakes (both natural and artificial) are slowly but surely getting depleted and polluted through sheer negligence.

Depletion

The 55 km long Swan river originating from the barren Shivalik hills in H.P. is fast turning the submountainous district of Una into a desert. The river, which continuously changes its route, is being regarded as *'Himachal's river of sorrow'*.

The river bed is now full with sand and sediments brought down from the Shivalik. With every downpour, the area under the desert increases. What makes the desertification problem much severe is that Swan has another 75 tributaries, which also plays havoc in the rainy season.

According to the document brought out by the Indian National Trust for Art and Cultural Heritage (INTACH) in 1989, indications are clear that the temporary phenomenon of metrological drought in India is tending to be converted into a permanent and pervasive phenomenon of desertfication undermining biological productivity or soil over large parts of the country (Anand, 1989).

Desertification often starts as a patchy destruction of productive land through inappropriate use. Since the large majority of people in India, have livelihoods based on land the long term decline of biological productivity of land, undermines livelihood and results in economic under- development.

It is primarily because of the lack of green cover on hills that there is progressive reduction in the supply of fresh-water in the hills and in the plains (Futehally, 1988). This is evident by the catchy phrase *"The health of the hills is the wealth of the plains"* coined at the March '88 seminar on the Palni Hills, T.N.

While controversy regarding the water depleting role of *Eucalyptus* and other foreign species continues, if a pragmatic view is taken, these trees have undoubtedly affected the water regime of hills calamitously.

As *Eucalyptus* grows, the ground water level goes down further year after year. *Eucalyptus* roots do not keep water together and also do not store it like the broad leaved the trees can do with their widely branching root system. An adult *Eucalyptus* tree dries upto 500 litres of water a day. Many mountain rivers in valleys and other places, which are still painted in old maps as water lines, today have not a drop of water. The use of chemical fertiliser in *Eucalyptus* plantations further creates the problem of poisoned water (Bahuguna, 1988).

Streams and springs which supplied copious water during past decades are now drying up, and Berjan Lake in H.P. dried up almost completely for the first time last year.

By and by taking the country as a whole, India receives more rainfall, than most countries, but droughts and floods can be attributed entirely to the lack of vegetation on land.

Unscientific limestone quarrying in the Mussoorie hills and indiscriminate commercial exploitation of forests in the area, are responsible for increasing environmental pollution in the Dun Valley and depletion of water resources. If more industries are set up, demand for the already scarce water is bound to increase.

After the much-publicised *"death"* of Mussoorie hills, it is now the turn of hilly district of Sirmur in Himachal Pradesh. In the name of development, the barren Shivalik hills are being ruthlessly blasted for extracting limestone. Scientists at the regional research station of Agricultural University at Dhaulakuan, on the basis of a two year survey, have concluded that the dust from plants does not allow pod formation in gram and linseed. The dust has also settled on the precious Sal trees near Rajbans (H.P.) restricting their growth.

Shivalik hill ranges are known to comprise unstable and immature rock formations. Blasting with dynamite and other ways of mining can play havoc with the hills. Besides with the extraction of limestone, even drinking water is likely to become scarce in the district. Already H.P. is faced with the gigantic task of supplying safe drinking water to its villages.

About 50 limestone and sand quarries in the 10 km long Saproon Valley (H.P.), spread along the Solan-Subathu road, are proving a big threat to the natural wealth and environment in the area.

The quarry had caused flash floods a number of times and had washed away some houses, animals and top soil thus denuding the area of vegetation. This debris keeps flowing downhill during the rains, resulting in road blocking and destruction of crops. In the valley the only source of wâter are natural springs and *'kools'* located in the mountain ranges. But blasting and debris are disturbing these natural sources of water. A natural spring at Sheel village has dried up. The people are worried that if quarrying goes on unchecked or is increased, they might soon have to fetch water from distant places. Besides, it, the reckless, quarrying for slate in the Dhauladhar mountains in the Chamba district of H. P. is a serious threat over the survival of the numerous major hydro electric power projects in the Himachal Pradesh.

Planning

According to the document brought about by the Indian National Trust for Art and Cultural Heritage (INTACH), 1989, "Deforestation, drought and desertification; perceptions on a growing ecological crisis", there is an urgent need for people and planners to recognise that desertification arises from ignorance of natural processes of renewability of soil-water-vegetation system.

Describing drought and desertification as a *'silent crisis'* the report says if proper attention is not focussed on the problem of recurring drought without further delay, the situation in this part of world too may become as serious and disastrous as in Africa.

While scientific data relating to the rehabilitation of our renewable resources—air, water, soil, flora and fauna is adequate, what is not satisfactory is the set up of the administration at district level, where the action lies. Apart from the fact that there is considerable *'resource illiteracy'* within the administration, 'the rapid turnover of officials, their lack of continuity creates serious problems.....' In this connection, the speech by Mr. B. Vijayaraghavan, Chairman of the T.N. Electricity Board, stressed the need for a completely autonomous Hill Development Council with an independent staff under the total, control of the chairman. Unless this is done and if the executive functions are left to the existing departments of Government, the same problems about lack of coordination and non-cooperation within the Government would continue.

Environmentalists from seven South Asian countries participating in a three day seminar (1989) on regional cooperation for protection environment proposed an integrated approach for optimum use of biological and economic resources, measures to tackle the pollution of three major river—the Indus, the Ganga and the Brahmaputra. Desertification and degradation in the Himalaya were cited as areas for regional cooperation.

Mr. Abul Ahsan (1989), secretary general *'SAARC'* secretariat, Kathmandu has said that shortage of drinking water in India in one of the issues that could be tackled from the south Asian regional cooperation angle.

Dr. Akmal Hussain (1989), a member of Pakistan's Prime Minister Ms. Benazir Bhutto's advisory group on economic policy, *'SAARC'* said there was an urgent need for devising new developmental methods to tackle common problems like poverty, Watershed Management and treatment of industrial wastes.

Conservation

Defence against drought and desertification is not always provided enhanced tree cover according to INTACH report (1989).

The most fundamental misconception, it says in drought management is the assumption that any general scarcity of water, either as surface run-off drought, soil moisture drought and acquifer drought, which are associated with desertification are the same as metrological drought or scarcity of moisture for plant, animal and human life is not necessarily connected with 'scarcity of rainfall'. It is man-made scarcity often associated with misguided attempts at development in agriculture, forestry or water resources. Most official water development projects in drought-prone regions fail to see groundwater shortages as ultimately needing to be recharged at the surface.

The fast depletion of the country's ground water sources due to over-exploitation has also engaged in the attention of the National Drinking Water Mission. It is feared that over exploitation of ground water resources could even hamper the Union Government's minor irrigation plans. Thus massive social awareness is as essential as the need for appropriate legislative measures in all states to check the over-exploitation of ground water resources.

The only way of utilising some of the vast run-off of the country's rivers is through natural recharging of acquifers. On a small part of the run-off is stored in surface reservoirs and tanks and the rest flows down into the sea. The wastage could be substantially reduced by restoring the natural eco-system which has been rudely disturbed in the process of 'development'.

Worthy of scrutiny in this context is the Union Government's Hill Reservoir Scheme, which laid emphasis on promoting more intensive watershed development through systematic water harvesting in the micro watersheds.

A voluntary organisation of scientific experts, 'Doon Science Forum' (1988), has come forward with a novel suggestion for meeting the increasing demand for potable water in Dehra Dun. It favours construction of an artificial reservoir, measuring 6 to 10 sq. km. below the Mussoorie hills where the average annual rainfall is 2000 mm. This multipurpose reservoir can meet the requirement of potable water and generate power to the tune of 2,25,000 units per day.

The toxic effects of lime-based industries are destroying the natural environment of the Forest Reserch Institute, Dehra Dun. Knowing the extent of damage it causes to ecology, certain district officials admit that quarries in the area should either be banned or improved scientifically. In addition, limestone is known to be a good conservator for water. If the entire limestone is taken out from the hills, even drinking water is likely to become scarce in the Dehra Dun district.

But the possibilities seem remote as most quarry owners have strong political links and the officials prefer not to enforce various laws.

Thus Management of drought calls for immediate attention and launching of programmes for *social forestry* and wasteland development through an understanding of inter-disciplinary knowledge of the relationship between soil, water and vegetation.

SOIL

Depletion

The Himalaya are the youngest as well as the highest and longest mountain range on the earth. They are still inching upward from the impact of India with the rest of Asia, and are naturally being weathered away, disappearing into the sea.

Tonnes of soil are being stripped from every acre of the mountains during each monsoon destroying the land on which Himalayan people depend for their livelihood.

The erosion of the Himalaya has greatly increased flooding in the Ganga basin, one of the world's most densely populated and important food growing areas. In all 300 million people are affected by what goes on the mountains (Lean, 1986).

The trees bound the soil to the mountain slopes and enabled it to soak up the fierce monsoon rains releasing them gradually to enrich the whole region.

Now the rains are a curse, not a blessing. Each year they wash away more than 12 tonnes of soil from every acre of the bare hillside. In worst areas 80 tonnes of soil are ripped from each acre. (Lean,1986).

Villagers watch all night for, landslides during the monsoon. Little wonder—20,000 landslides have been recorded in a single day in 1986 sweeping away the terraces of fields curved laboriously into the mountains side and burying whole villages and their inhabitants. (Lean, 1986).

Landslides, soil erosion and silting of water sources is enough to show of the reckless abuse of the forest and natural wealth for which Shimla (H.P.) was once, known for. With the rapid deterioration in its environment Shimla. The *'Queen of Hills'* is probably the only place in the country which is warming up like a*'green house'* (Sharma. D, 1988).

Excessive discharge of smoke along with the increase of CO_2 content in the air is forming a dense layer in the atmosphere which does not allow the resulting heat to escape. Known popularly as the *'green house effect'*, the phenomenon is causing an alarming shift in the microclimate of Shimla especially during winters. The smoke in the atmosphere comes down along with the rains in the form of an acid which has an adverse effect on the vegetation. Thus the increase in the acid content in soil and in the atmosphere is perhaps the main cause behind the change in the vegetable habitant. Much of soil degradation has been noticed at Chakkar, Bharari and Sanjoli areas.

Startling graphs exhibiting the loss of forest land, topsoil, stratospheric ozone, and species all follow the same pattern of sudden, acceleration in the latter half of the 20th century. If thought why these climatic changes are taking place, it is because the human population is surging because the industrial scientfic and technological revolution magnify the environmental impact of these increases, and because we tolerate self destructive behaviour and environmental vandalism on a global scale.

The impact of ecological degradation in Sikkim Himalaya is noticed by landslides and avalanches, which of course are not novel events, but what is new is the alarming increased frequency with which they occur and the human toll they exact these days.

Each year the monsoon ravages the deforested slopes and washes away millions of tonnes of topsoil, and causes floods with heavy losses of life and property in the plain and lowlands, (Nadkarni 1990).

As the soil goes, crops fail. Rice yields have dropped by a fifth in just five years in the hills, maize yields by a third. Thus geological processes are being vastly aggravated by man's struggle to survive. Leading members of the tribes are afraid, that the land around the villages will become deserts. Sir Edmund Hillary (1990) calls Himalaya *"an ecological slum"*.

The population is growing rapidly and as it does more and more farmers try to carve fields out of steeper and steeper slopes exposing them to erosion. Besides this, there are as many cattle as people and they graze the land nine times more intensively than it can take, tearing up the grass, crippling trees, and eating seedlings before they can grow.

As wood becomes scarce, people stop using animal dung and crop residue as fertilizers and start using them as fuel. And as this happens the land already fragile from constant cultivation quickly loses its fertility. As crop yields fall, erosion speeds up and productivity drops off dramatically.

Thus Government should concentrate on a less dramatic but more insidious kind of desertification—the long term degradation of soil and vegetation in dryland areas used for farming or grazing where there are serious problems of poor land management, heavy population pressures and where farmers often have to resort to environmentally destructive farming techniques.

Planning

The green house effect and the stratospheric ozone depletion fit the profile of international security issues of global significance. These phenomena certainly will, intime, produce effects big enough to threaten international order, (Gore, 1989).

Scientists, foresters, environmentalists and government officials are seriously exploring the feasibility of an ambitious long term enterprise planting enough trees around the world to ease the threat of global warning, (Hindu, 24.9.89).

Nelson (1988) a natural resources specialist in the World Bank's Environment Department says, "we need to find profitable farm system involving trees that farmers will have an incentive to use, and we need to promote economic policies that increase these countries ". He adds, "But the fact is that we currently have very few of these kinds of technologies to offer the farmers in most of these dry areas. We are still looking for answers."

However, poineering projects by the U.N. agencies have shown that land can be saved if the land is left alone for a year or two.Trees and grass grows rapidly stabilising the soil and cutting erosion by 90%. If grass is then cut, it feeds four times as many cattle as would live by grazing the area. Planting fruit trees saves the soil and provide the farmers with produce.

Farmers at first suspicious of the project are now anxious to give them land to rehabilitate. U. N. Aid agencies are ready to provide money to help Himalayan people tackle its crisis.

The country's slow moving and labyrinthine bureaucracy has been unable so far to come up with co-ordination programme to save the soil.

Conservation

The World Bank is watching an experiment being carried out in parts of India of planting rows of *Vetiver grass* to help conserve soil moisture.

Experiments by scientists at the Central Road Research Institute, India (CRRI) have proved that growing *Vetiver grasses* and bushes on the hill slopes can prevent landslides that have over the years claimed human lives and cattleheads besides damaging properties along hillranges.

Mr. Jagannatha Rao (1989) head of the geo-technical engineering division of the CRRI, prived that the erosion problem could be effectively checked by promoting vegetation growth on the denuded slopes. He has suggested, that methods like the laying of *"Geo-textile"* jute mesh, coir mesh, geogrid mesh or spraying of asphalt mulch on the exposed slopes could act as promoters of the vegetation growth on the denuded slopes. The institute had succeeded in preventing landslides in an area of 6000 sq.mt. in Western

Himalaya and Nilgiri hillslopes so far by adopting several technologies. Rao said studies had also shown that surfacial erosion slides could be controlled by simple techniques, involving the use of coirnetting, jute netting or asphalt mulch in conjunction with vegetative covering.

The Institute of Engineers (India) in the Seminar (11, 12th July, 90) on "Soil improvement by *"Geo-textile"*, stone columns and chemical grounding" has given a cell for wide use of geotextile to check soil erosion. The engineers have cited the success of geotextiles in preventing constant soil erosion along the river bank along the Calcutta and Haldia ports, (Ray, 1990).

Mr. G.C. Naik , Former Joint Director of Agriculture Govt. of Orissa, has favoured *Khus* grass as a mean of vegetative soil and moisture conservation (1990) . According to him, the vegetative system of soil and moisture conservation using *Khus* is cheaper than engineering mechanical earth bunds, and more effective since no maintenance cost are involved. In Orissa, farmers had been using it for many years for field boundaries although its technology for soil and moisture conservation was quite unknown.

Khus planted on contours immediately after the first monsoon showers seldom required any irrigation. It withstood the summer hear and came up vigorously in the second monsoon season to form a hedge. Further, the earthen bunds impound rain water for a day or two in light soils in the uplands and for five to six days in heavy soils.

The vegetative system of soil and moisture conservation is dynamic, effective, low cost and the appropriate technology for farmers. Once planted the crop needs no further care or maintenance.

There is validity in the point that *Khus* survives best in high moisture areas, but there is also no doubt that it can grow in dry soils if planted just prior to the monsoon season. The World Bank prescription is in fact, an indigenous Indian Plant which has been adopted by farmers.

FORESTS

India is a country with extremely diverse ecosystems within the same country, there is the hot desert of Rajasthan, the cold desert of Ladakh and some of the highest rainfall areas in the world—Chirapunchee. So there is no clear picture regarding the unique ecosystem of the region and reliable data on its forests, waste lands, arable land and other land resources, (Agarwal and Narain, 1990).

The records of the forest departments show 60-65% land under forest cover. The landscape imageries indicate that the forest cover in U.P. Himalaya is only on 28.7% land. Actual forests with more than 60% , crown cover are only on 4.4% land area. The rest 15% and 9.3% land areas are under moderate and poor forest cover, respectively. Thus there is a huge gap between reality and what is being projected.

According to the figures of the forest department the maximum area in the U.P. hills is under Oak forest (4036 sq. kms) followed by pine (3990 sq.km.) and Sal (1051 sq. km).

It is thus evident that the present state of U.P. Himalaya is sombre. The dissenting voices from the communities dwelling in the region and large scale migration are further indicators of a fast approaching ecological collapse. Firm remedial action does not brook delay, (Singh & Saxena, 1988).

With the plundering of the ecosystem in the hills the mighty mountains of the Western Himalayan zone, comprising Western U. P., J&K. and H.P., have begin to move, (Sharma, D., 1987).

India may have already gone beyond the threshold of tolerable levels in the loss of its forests said Mr. Attila Karaosmanoglu, the World Bank Vice President for Asia in the year 1989.

Those travelling from Kalka to Shimla after a decade or more are appallel to observe the *'naked'* hills en-route. One observes only stumps all around as one derives from Manali towards the Rohtang pass. That wantom destruction of forests in H.P. has resulted in barren tracts of land. The sources of water have gone dry following the clearing of forests (Sharma, N.D. 1988). Pastures have virtually disappeared causing shortage of fodder for the cattle.

A number of environmental factors are responsible for triggering a series of avalanches, which is a natural phenomenon in the high hills. Denudation of the forest cover is further aggravating the problem.

Mr. Karaosmanglu said in 1989, In India "Population is the major cause of deforestation and degradation of watersheds. An important aspect of the forest depletion problem, he said, was that on publicly owned forest land, private concessionaries, typically paid low user fees to the Government, with a flat charge based on volume of timber removed, serving for destructive logging practices so that forests were "mined" not harvested. People complain of involvement of the forest staff in illict felling of trees and smuggling of timber. Most of the "irregularities" are being committed in the name of private sale.

Several forest fires are raging in diferent parts of H. P. causing large scale damage to the precious forest wealth. This has caused extensive damage to the trees besides causing loss of valuable resin in the pine forests. According to fire officials the main cause of the forest fires is the hot dry weather.

The high velocity summer winds, lack of preventive measures like "fire lines" and absence of rains are helping these fires to advance to neighbouring areas.

Air pollution is contributing to the decline and ultimate death of forest trees.

Air pollutants injure trees both directly through foliar damage and indirectly by leaching nutrients from the trees and soils.

As air pollution's effects on the ecosystem become progressively more severe, the whole ecosystem can deteriorate.

Because of the inherent biological and chemical complexity of forest declines, the exact mechanisms leading to the damage and morality of some tree species have not been determined.

At high elevations where the trees are declining, the acidity of the cloud moisture is about 10 times greater than at low elevations and 100 times greater than that of unpolluted precipitation.

Planning

India faces an extra ordinary challenge over the next few decades. The population is growing rapidly and the dependance of the majority of the people on a biomass-based subsistence economy continues. The country's land and water resources are coming under increasing pressures will completely destroy India's ecological systems unless it can soon find highly productive systems which are at the same time ecologically-sound and sustainable. This will pose a major scientific, social and political challenge for India.

The best starting point for environmental management is often water and not trees. Villagers are usually desperate for irrigation water and, therefore, keen to develop local water harvesting systems. Once they get a water harvesting system, the next step—to protect the catchment and bring it under grasses and trees—comes naturally to the community. But it would require rethinking not only about the current structure of the India's state but also the current percepts for rural planning.

India being a country of extremely diverse ecosystem, the same kind of land use and the same kind of biomass production is not possible in all these ecosystems on a sustainable basis. Rural development planning must, therefore become *ecosystem specific*.

But to involve the people, the *action level* will have to be the human settlement. Ecosystem-specific rural development guidelines can inform and guide village level plans and projects.

Conservation

Today, an environmental disaster in the Himalaya has become a subject of hot debate all over the country. The Himalayan development is being given high priority in national planning and lot of money is being allocated for its development.

The very first measure to stop the spread of wasteland is to undertake an intensive plantation programme on such lands. To save and preserve the existing forest areas in an even more challenging task. Development of alternative sources of energy should be the first step towards saving present forests.

Only those species should be planted which are native to the place and which also fulfill the day-to-day requirements of the people. Broad-leaved fodder trees, palatable and nutritive grasses and leguminous fodder species should also be included under the plantation programme, (Singh & Saxena,1988).

Village ecosystem planning i.e., integrated planning for the croplands grazing lands, forest lands livestocks and water systems of a village—can be attempted only at the village level, and by the villagers themselves.

Given the right institutional and legal frame-work for community action, villagers can manage their village resources on the basis of the principles.

The government has to play the role of educator, trainer and provider of technical assistance. Researches will also have to be reorganised. Modern science and technology must build upon the social and ecological foundations of traditional knowledge systems of land use, water harvesting, cropping system and agroforestry.

Reforesting the earth can be part of the remedy to slow the global warning underway because of the *greenhouse effect.*

The Chipko movement leader, Mr. Sunderlal Bahuguna is of the opinion that the ideal way to promote regreening of the hills will be to allow the indigenous shrubs and broad leaf trees to regenerate from residual root stock.

Under social forestry programmes people should be given species that will yield fruits and fodder protection against biotic interference will result in revegetation, he says.

Aerial seeding for afforestation of degraded lands is one of the Government's projects, (Chaturvedi,1988).

The desert development programme is ushering in a silent green revolution in the rocky and cold deserts of H.P. by dotting the brown landscape with greenery and bringing economic prosperity for the residents. The emphasis under the programme is on afforestation and irrigation is Spiti sub-division of Lahual-Spiti district and the Pooh sub-division of Kinnaur district (Pubby,1986) .

In a small, humble way, the programme for greening of the Western Himalaya has been launched in the Hemal Valley of Pauri Garhwal, U.P. In this programme, instead of an afforestation project prepared by top bureaucrats, the areas to be afforested, the selection of trees species as well as the rates of payment for various jobs will be worked out by the Mahila Mandal themselves, (Rai, 1989).

Much of the blame for the destruction of forests rests on an outdated and unreal forest policy, introduced in 1952, which failed to recognize the ecological significance of forests. The long awaited new forest policy makes it clear that forest based industries will no longer be allowed to plunder the country's forests. It says that the practice of supplying forest produce to industry at concessional prices should cease. The schemes and projects which interfere with forests in catchment areas and geologically unstable terrain should be severly restricted.

The U.P. Pollution Control Board has taken measures to implement the ban on polluting industries in the Doon Valley, (Painuli, 1989).

Barries zones of *Cedar* and pine trees should be created all around the mining areas in the Himalaya to arrest aerial spread of dust," Plants around which spiders spin their webs shoul be planted at mining sites so that the web can arrest the dust and reduce, air pollution to same extent.

TREKKING

To reduce the negative environmental impact and for getting away from the everyday whirl of living, enjoying the bubbling streams, quiet forests paths, and lonely mountain spaces, the coinage of a novel activity in the Himalaya has come into vogue-development of *"Adventure tourism"*.

The mark of the beginning in this field is evident from the access to road in remote areas that has greatly improved. The quality and the number of airfields in the mountains have multiplied. Nowadays, tens of thousands of people are walking the hill tracks, floating down the rivers driving over the great Himalayan passes and landing in tiny airfields in remote valleys.

The joy of adventure tourism, to pit one's skill and endurance against nature's roughest of rock and snowy heights has begun to gain much popularism today. Spring comes to the Himalaya in the late March. The deep and silent jungles along the 2500 km long Himalaya, during this season the annual assault will begin. Thousands of tourists, trekkers and mountaineers swarm over this young and fragile mountain world by the second week of April, (John, W. 1990).

Every year, more than 75 lakh people visit the Indian Himalaya. Their number in 1961 was about 20 lakh and 50 lakh in 1980. Though it is a very conservative estimate, the specific example of Ladakh will bring home the dimension of their tourist invasion. In 1975 there were hardly 550 tourists but in 1985 they numbered 7000. When Sir Hillary and Tenzing Norgay climbed Everest in May 1953 there were less than a dozon mountaineering expeditions and less than a hundred trekkers. Now about one lakh trekkers and 3000 expeditions visit the Himalaya in a year, (John, 1990).

The Himalayan Club founded in India in 1928, assists virtually all climbing parties across, the entire range. Further the Indian Mountaineering Foundation 1958, supports and co-ordinates mountaineering expeditions and other related activities.

In 1931, the British mountaineer Frank Symthe returning from his Kamet expedition discovered in the upper reaches of Garhwal Himalaya 3600 m above sea level an alpine valley with the difference. Here slopes, ledges, terraces, barren rocks, running water coarse grit were all crowded with millions of flowers of numerous hues and descriptions calling it fondly "the Valley of Flowers" —a valley of peace and perfect beauty where the human spirit may find repose. Every year from June to September numerous enthusiasts trek 14 km from Govindghat to Ghangharia in the Chamoli district of U.P. A stiff uphill climb for another 3 km from Ghangharia across the turbulent Pushpavati river and glacier leads to a humble wooden bridge from where the first view of the magnificent valley sprawling in the foothills of the lofty Rataban peak 20231 ft. can be had. Every season around 800-1500 tourist on an average visit the valley—a major section of this crowd are Sikh pilgrims to Hemkund Sahib located at 4650 m.

Depletion

The world's highest mountain range is today under threat from mass tourism and excessive adventure activities. The number of visitors have caused extensive and often irreparable damage to the delicate*"Himalayan Environment"* They trample down high meadows, axe trees in valleys, disturb wildlife, uproot plants, pollute rivers and springs and leave behind a long trail of garbage even in remote corners of the Himalaya, thereby putting severe pressure on local water supply, sanitation facilities, food availability and the traditional way of life in general. Landslides and soil erosion have increased and mountain air, that fresh and fragnant air is now thick with petrol and diesel fumes, (Rani, 1989).

This is evident by the devastation that has taken place in the Nanda Devi Sanctuary, one of the inaccessible areas of the Himlaya, lying deep inside the Garhwal Himalaya, beyond Badrinath, is the last village on Indo-Tibetan border, where only a few hardy trekkers and mountaineers could reach. In 1934, Eric Shipton and Bill Tilman found a way through the Rishi Gorge thus opened the flood gates for hoards of adventurers. Within few years the rich coniferous and birch forests at several camping sites were wiped out, rare wildlife species became endangered, landslides became frequent and sheep which carried goods devoured flowering plants, shrubs and more importantly, the grass which stabilises the weak mountain walls. Due to the destruction the entry of visitors was banned by the Government in early 80's. In 1988, the sanctuary was declared a biosphere.

"Tourists and environment are generally in disagreement" writes Dr.Tejvir Singh , the Former Director of the Instt. of Himalayan Studies and Regional Development HNB Garhwal University as several areas in Himalaya are considered *'critical'* by mountaineers and environmentalists. They warn that pilgrim centres and popular trekking destinations like Pindari glaciers, Dodi Tal in U.P., the Everest base camp and Anapurma Sanctuary circuit trail in Nepal and the Karokoram region in Pakistan are threatened by human intrusion.

The Himalaya is made up generally of weak deformed rocks, still in the process of formation and is easily affected by rain, shocks of explosions earthquakes and vibrations generated by heavy vehicles and human interferences. Beauty spots can be literally worn away by road, traffic shattered by noise and burried beneath litter.

The British Mountaineer Frank Smythe writes that "Growing population cattle heads excessive grazing initially and now tourist traffic in short due to increased human interference, the beauty spot of Garhwal is dwindling".

Exploring nature's garden one comes to grip with the impending crisis in the *'Valley of Flowers '*. Frequent human contacts seems to be threatening the very existence of the valley. As it is the valley is faced with other threats. As a result of Global changes in temperature, the glaciers are retreating the ever increasing warmth is becoming conductive to the growth of taller species of plants. Though the number of visitors has been steadily increasing but their quality is not good.

"Disuse has its own destructive effects". It is quiet evident by the fact that the visitors don't walk across laid out paths, they pluck flowers, uproot plants and this depredation has caused havoc. Almost every visitor plucks handful either as an offering or as souvenir. But by the end of the trek they wilt away and are discarded by the nature lovers.

Thus the existing situation seems to be hopeless. The callous attitude of the visitors and the authorities is bound to take a heavy toll of the valley.

Planning and Conservation

The position today is that in the Himalaya, as in most other wilderness areas of the world, visitor impact both in terms of density and spread has created serious problems of degradation of the environment and thus there is need for immediate attention and corrective action.

In this respect, Sir Edmund Hillary and Capt. Mohan Kohli (1990) have evolved the plan for the formation of the*"Himalayan Adventure Trust"* to mobilise the support from the mountaineers, trekkers, to protect the Himalayan environment, its flora, and fauna and natural resources to evolove necessary guidelines in consultation with the Himalayan countries concerning adventure tourism to avoid over grazing of certain popular trails, and achieve a fair spread of trekkers and mountaineer throughout the Himalaya.

Assessing the visitor impact on the Himalaya, Mr. Sudhir Sahi a veteran mountaineer and the secretary of the organising committee of the first International Meet on Himalayan Environment which took place on March 30-31st, 1990 in New Delhi, observes, "Once the carrying capacity i.e. the number of visitors absorbed in an area without adversely affecting its environment and local population is assessed, the number of visitors can be regulated".

Such assessment of the Himalayan environment will take at least a few years but the present situation demands immediate solutions like the following:

(a) The entry of trekkers and mountainers should be either banned or restricted in all critical zones.

(b) Environmentally critical areas like Gangotri-Gomukh regions should be declared wilderness area and entry should be regulated.

(c) A levy of $ 5 and Rs. 25 should be charged from trekkers and moun taineers as conservation charges as being done in the Annapurna Sanctuary area in Nepal. The money thus collected should be ploughed back into local territory.

(d) The number of hotels and rest houses should be restricted and no new construction should be allowed in heavily crowded hill stations like Nainital, Mussoorie or Shimla.

(e) No buildings should be allowed beyond a certain altitude (construc tion of concrete structure in the high altitude Hemkund lake near the famed valley of flowers and Gangotri region has damaged these unique places).

(f) Kerosene or LPG should be provided at hill stations and along the trekking trails to stop the felling of trees.

(g) Trekkers and mountaineers should be asked to bury or burn the garbage.

(h) Efforts should be made to use biodegradable packing material.

(i) Suitable toilet facilities should be constructed at hill stations and popular trekking trails.

(j) Special expeditions should be organised to clean garbage trails.

(k) Facts about Himalayan condition should form a part of the *national environmental awareness campaign* and

(l) The local population should be involved in all conservation project in the Himalaya.

However, many clubs and association have been conducting an ongoing series of educational and familiarisation programmes, thus it is possible to maintain and disseminate updated information on the status of the *"Himalayan environment"*.

Thus we must prevent the wanton destruction of all that is fresh and beautiful and enjoy to the full, the superb beauty o f nature in the remote mountain areas.

REFERENCES

Agarwal, A., & Narain, S. 1990. Towards Green Villages, Hindustan Times.

Ahsan, A., 1989. 'SAARC' Cell to Protect Ecology, New Delhi, Feb.1989.

Anand, V.E., 1989. Plight of Waterless Villages, Indian Express.

Bahuguna, S.L.,1988. Perils of *Eucalyptus* and *Acacia*, Indian Express.

Chaturvedi, A.N., 1988. Sowing Forests from the Air, Indian Express.

Futehally, Z., 1988. The Hills are Alive-Just, Indian Express.

Gore, Jr. A., 1989. Ecological neglect will threaten world order—Euro space Symposium, Paris, May 1989.

Hillary, E, Sir., 1990. The Crumbing Himalayas—Conference on Environmental Protection of Himalayas, New Delhi, March 1990.

Hindu, 1989. Trees to halt climatic change, The Hindu, Madras.

Hussain, A., 1989. 'SAARC' Cell to Protect Ecology, New Delhi, Feb. 1989.

INTACH 1989. Indian National Trust for Art and Cultural Heritage–Studies on Ecology and Sustain able Development Report, March, 1989.

John, W., 1990. Trampling the Himalayas, Hindustan Times.

Lean, G. 1986., Himalayas are being Washed Away, Times of India.

Nadkarni, V., 1990. Environment–Key to Survival, Times of India.

Naik, G.C., 1989. Vote for *Vetiver*, Indian Express.

Nelson, R., 1988. World Bank Funds to Conserve Soil, Indian Express.

Painuli, P., 1989. Stern Steps Against Pollutant Units, Times of India.

Pubby, V., 1986. Forests Gain Ground in H.P., Indian Express.

Rai, U., 1989. How Green will Our Valleys be, Times of India.

Rao, J., 1989. Grasses to Prevent soil Erosion–Central Road Research Institute, India.

Ray, A.C., 1990. Seminar on "Soil Improvement by Geotextile", Institute of Engineers, Calcutta, July 1990.

Sharma, D., 1987. Western Himalayas Moving, Indian Express.

Sharma, D., 1988. Queen of Hills is Warning Up, Indian Express.

Sharma, N.D., 1988. Forest Depletion Continues, Indian Express.

Singh, V. & Saxena, A.K., 1988. Wasteland in U.P. Himalaya., Times of India.

Varsha Rani, 1989. Valley with a Difference, Hindustan Times.

Part III
Case Studies

9

THE PROPOSED FUNCTIONING OF THE U.P. HILL DEVELOPMENT CORPORATION

S.P. Nautiyal

Introduction

The eight hill districts of Uttar Pradesh cover an area of about 55,000 sq. kms. and have a population of about 4.4 millions. On account of distant location and difficult terrain the hill areas have not been developed to the extent the other areas of the State have. The industrial backwardness is evident from the exceptionally low ratio of deposits to investment from the Banks in the hill districts. For the last many decades no serious efforts had been made and as a result of which large scale exodus of man-power from the hills has upset the economic balance. It can be restored by speeding up the development of the hill areas, agriculturally, economically and industrially.

The Uttar Pradesh Hill Development Corporation was established in 1971 as a State Undertaking. The authorised capital of Corporation is two crores rupees and on 30th November 1975, the paid up capital stood at 03.4 lakhs rupees. The Memorandum of Association of the Corporation, duly approved by the Government, of Uttar Pradesh, clearly defines the aims and objectives which are very comprehensive. The Corporation shall aid, assist, promote, establish industries of various types singly or jointly, under take marketing, which is the most important bottleneck in industrial development, and render advice to the Government, private or other Corporations working in the area. It is also aimed to undertake purchase and sale of various types of goods for proper distribution and price control. For carrying out the comprehensive task, the share capital, organisational structure and powers of Hill Development Corporation are very inadequate.

How to Achieve the Aims

The Corporation has been in existence since 1971 but had no systematic plan of work and, therefore, it has not been able to create an impact in masses. Appendix 'A' gives summary of the work so far carried out from which it is evident that these were taken up in scattered manner.

The Board of Directors of the Corporation in its meeting held on 12th September, 1975 approved the following organisational pattern:

Chairman

Managing Director

Experts

Marketing | Forest | Tourism, Transport & publicity | Agriculture, dairy, Animal Husbandry | Mineral

Engineering & tools | Horticulture

District Managers

Uttar Kashi | Tehri | Dehradun | Chamoli | Garhwal

Nainital | Pithoragarh | Almora

The Board has also approved appointment of District Managers at each district headquarters which will enable decentralization of the powers and functions of expedient supervision, and execution of projects. The organisational get up will enable to implement the plans and the schemes with the help of the officers of its own cadre instead of getting people on deputation who do not have much stake and involvement in the effective working of the Corporation with commercial orientation. Technical assistance, however, will be drew from various government departments on short term basis.

Functioning of the Corporation

The Hill Development Corporation is engaged at the moment for creating suitable organisational pattern to prepare schemes for the entire area itemwise, their evaluation, financial analysis, and assessment of man power requirement. The Corporation is making schemes which can be implemented in the near future, agro-based, horticultural and forest based. Rest of the schemes will be arranged according to their gestation period.

The functions of the main body or apex body wiil be to draw master plans data collection, marketing, general overseeing, procurement of finances and budgeting. In order to prepare master plans the staff at the disposal of the Corporation is not adequate, but in order to keep the salary and wages bill low, it is proposed to induct experienced retired hands on honorarium basis on short term basis. This will not only save money to the Corporation but make available expertise at nominal cost.

The District Managers would be principal officers who would implement and execute work in the districts. They would also send indepenedently plans on the basis of their knowledge of their areas to the main office. It is envisaged that the district office would take up full work in the district and will, therefore, necessarily enlarge fast providing employment to the people of each district. Creation of District Manager's post would give equal opportunities to all the districts, for both employment and also involvement in the developmental schemes.

Thus stationing of the main office would be immaterial from the point of view of employment. What is most important is that a bird's eye view of the possible development in stages should be carried out at the earliest so as to create impact amidst the hill people. Assessment of the work load, manpower requirement, opening avenues for employment to the unemployed educated, financial requirement must be made by the Hill Development Corporation's main body.

Relation with Other Organisations

The Government of Uttar Pradesh are spending vast sums on the development programmes through developmental agencies and other Corporations whose responsibilities are not only towards the hill areas but to the entire State. Last year total amount spent in developmental programmes was about Rs. 22 crores and similar amount is proposed to be spent during the present year. The total plan outlay for the Fifth Plan is about Rs. 300 crores. Since the Government has created the Hill Development Corporation to concentrate on the economic problems of the hill people, it appears necessary that the apex body for coordination of the developmental programmes in the hills should this Corporation. The aims of the Government Departments and the various agencies including Hill Development Corporation are identical and so it would be necessary to work in cooperation rather than in isolation. There is so much to be done that it can be achieved only through the combined efforts of all concerned.

In addition to the Government Departments and the Corporations there exists a large number of private enterpreneurs with different financial status. It is necessary to bring these people in the fold of the Hill Development Corporation with a view to utilise all available expertise, experience and financial resources. Such coordination can easily be worked out in large industries based on the forest, mineral, transport, tourism, dairy, tool making, horticulture, wool etc. It is also necessary to properly define the functions of the Government Departments vis-a-vis Hill Development Corporation to avoid any duplication. It is of paramount importance that the attitude of the Government Departments should not be authoritarian and coersive but it should be corrective and educative. In any corporate venture the technical skill and expertise is built up gradually which is achieved through proper training and education. Any coercive attitude at the beginning would be harmful. In fact many cooperative ventures die a premature death on this account. In the scheme of alround development of the hills each Government Department should take into account the need of the venture. This can be illustrated as follows:

(a) The Forest Department shall make it convenient to allocate suitable raw material at the nearest-point of the venture and in case it is not available in adequate quantity in the area, it will take immediate steps to introduce plantations to keep the industries running for long periods. It is known that a useful timber takes long time to reach exploitation stage but attempts should be made without delay.

(b) The Public Works Departments shall take note of the need of roads in the particular area. The Hydel Department should make available power either through grid or by setting up micro-hydel stations.

(c) The Horticulture Department should create nursery facilities at different points and ensure that appropriate type of samplings are supplied to the various people. They should also ensure proper supply of pesticides and fertilisers, at proper time. Fruit trees take 5-10 years to yield and, therefore, any laxity at this stage is bound to result in heavy loss in terms of money and time.

(d) Animal Husbandry Department should make phased programme of improving cattle, and poultry breeds and also supply pedigree stock at convenient points.

(e) To avoid duplication and overlapping and to save expenses, the various Corporations working in the hills should meet once in a year or 6 months for discussing various aspects of development. The schemes prepared by the Hill Development Corporation should be discussed with the representatives of the concerning departments of the Government and other connected bodies. These can be discussed by committees headed by the Minister-in-Charge of the subject. Once the schemes have so been vetted and finally approved by the Board, there should be no interference from any source.

Working Pattern

The compact organisational pattern approved by the Board is expected to be effective machinery for implementing and guiding various business ventures. It is intended that each such venture shall be autonomous subsidiary unit of the Hill Development Corporation. The schemes will be drawn on the basis of village level development and involvement of the people eliminating the middle man. This is expected to bring maximum financial benefit to the actual workers of the unit. This can be achieved only by making all concerned share holders so that they would consider the business as their own and thus would be deeply involved. The profit would be ploughed back after suitable deductions on account of capital investment and share of the Corporation on account of expenses. Since the main body of the Hill Development Corporation will be small it is expected that the share of the Corporation shall be limited to a nominal percentage. In short, the Hill Development Corporation will function as holding company and numerous autonomous units, subsidiary to the Corporation will be run by the people with its technical and financial assistance.

While setting up industries it shall be moral duty of the Hill Development Corporation not to step on the toes of the individual enterpreneurs of cooperative ventures. In the larger interest of the industries efforts will be made to persuade such units to participate with the Hill Development Corporation for systematic development, exploitation and avail of the marketing facilities that will be created by the Corporation.

Marketing

One of the serious hurdles in the development of the hill areas is lack of proper marketing facilities of the produce. A strong organisation under the Marketing Expert will have to be created for ensuring sale of produce, quicker realisation of money and providing for regular flow of commodities, thus, relieving the workers, growers and producers of marketing worry. The marketing agency will also be responsible for cold storage and transport of perishable goods etc. The organisation will also contact foreign countries for selling of the produce either through STC or by itself.

The fixation of price of various products is not intended to be done at the source. In the beginning national payments can be made to the producers just to meet their working expenses and maintenance. Final payment will be made after goods are sold and accounts worked out jointly with the shareholders. This will enable ploughing back of "profit" to the producers. In times of slump the produce can be sold at convenient price and save total loss to the producers.

It is proposed to establish marketing centres at rail heads viz. Tanakpur, Ramnagar, Kotdwara, Haldwani, Rishikesh and Dehradun for export oriented goods and at numerous points in the interior for sale and distribution in the areas.

To illustrate this, let us take fruits and potato and other perishable articles. A broad delineation of growing belts shall be made and divided into unit such as Jaunsar-Dhanaulti-Chamba, Chamoli, Guptkashi-Dhampur, Pauri area, Pithoragarh area, Ramgarh-Nainital, Almora area and Champawat area. After grading, the produce will be selected for:

(a) Local consumption either for preserving or crushing or direct consumption or production of citric acid, tataric acid, etc.

(b) Consumption in adjacent areas;

(c) Consumption in the larger towns, and

(d) Export to foreign countries.

The material so collected will need to be stored. Cold Storage facilities will be created by the Hill Development Corporation in order to release the produce in limited quantities for price control. The cold storage will be at the highest altitude, near highways for transport facilities and near power lines for economic running of the unit. It may also be convenient for the Marketing Division of the Hill Development Corporation to hold auctions at cold storage centres after fixing the minimum price. Thus, in each area the entire process of production, transport, processing, packing, storage and marketing will be considered as integrated unit to be manned by the people of the area and payments on account of salaries and wages etc. shall be made out of the earning of the business. Thus, the Hill Development Corporation shall not have to employ regular hands or feed a high salary and wage bill. The entire scheme with financial involvement of persons engaged in the business and the payment to be made to various categories of producers will have to be worked out. The management of the Sub-Corporations shall be done by a Board consisting of an employee of the Hill Development Corporations Chairman, and two Directors from the share-holders. This pattern will be followed for every item of production. The involvement will be so close that any laxity on the part of the individual would virtually be wiped out.

Setting up Industries

The grouping of industries shall be according to the traditions of the people to start with. Since education is rapidly spreading in the hills, more sophisticated industries could be set up in near future according to the availability of raw material climatic factors needed for setting up certain chemical and electronic industries, the ability of the industry to bear transportation impact and availability of power, manpower and finances. The aim of setting up industries would be to meet local consumption as well as to meet export requirements.

A man power assessment including, their distribution and qualifications will be made for deciding as to what type of industry should be established in a particular area. This is very vital for successful implementation of the schemes keeping in view right climate, terrain and workers.

A fairly large number of hill people are working in various industries elsewhere. They shall either be coopted with projects concerned or hired for short terms or to run the projects.

Training shall be provided against sanctioned schemes to the persons who shall be engaged in the business. This will result in maximum and best utilisation of the training given to an individual.

A few examples are given below:

Forest Based Industries

(i) Based on wood working:

Wood industry of different types would be set up at convenient places right from the source down along the timber exit routes. The logging of the areas earmarked for forest fellings will be undertaken on cooperative basis by the people of the area. The Forest Department would be approached to arrange such quantum of felling which may provide continuous trade to the people of the area. Introduction of modern methods i.e. use of power saw, mechanical haulage, aero ropeways and other mechanical appliances would make it easier to employ the educated unemployed in this field.

Manufacture of packing cases shall be taken up in accordance with the need of the area. As far as possible, manufacturing would be done in the vicinity of users instead of transportng the cases from distance places. This will also give part time employment to the local people of the area during the off season or leisure and provide opportunities for additional income. Saw mills are also proposed to be set up in these areas specially for higher classes of timbers, not titherto worked from distant areas. Units for processing of timber such as pressing, preservation, treatment etc. and integrated wood working units are also proposed to be set up at convenient places.

(ii) Other forest based industries

(a) Paper and pulp, match, wood wool etc. are other industries which need to be taken up. Proper investigation will be made with a view to finding out possibilities for establishing such units keeping in view the availability of raw material, transport facilities and market requirements.

(b) Investigation will be made regarding the utilisation of forest produce other than wood for various industries such as pine needles for making board, fibres for making paper and ropes, seeds for oil etc. and, if found feasible, units will be established, as far as possible, on cooperative basis to utilize such products.

Resin Collection

Resin collection will be done by the local people as share holders of the sub-corporations set up for the purpose and supervision will be done by the District Manager of the particular area. The final payment will be made only after sale of resin and thus it is expected that each worker of the subsidiary corporation will get much higher return compared to earning as a labour.

In the ensuing season it is intended to have areas allotted around Tilwara and Champawat where two turpentine factories of the Hill Development Corporation are being established. Areas of Uttarkashi andTehri districts have already been allotted by the Forest Department. It is expected that the needed 36,000 quintals of resin would be tapped by this system for the above factories during 1976. Perhaps, in due course of time the entire resin tapping can be carried on the same pattern by the Hill Development Corporation.

Mining

One of the principal mining industry which can be developed fast is the limestone from Mussoorie belt, where numerous miners are already engaged. The mining work today is being carried out in a very unscientific manner causing serious erosion problem and hazard to workers. The Hill Development Corporation should take up the task of creating a sub-corporation including the mine owners, the Mineral Development Corporation of Uttar Pradesh and the people as share holders. The mine owners in lieu of their capital involved shall be paid suitable dividends. The Mineral Development Corporation will provide technical assistance by way of mining engineers and by providing training facilities to local people in surveying, supervision etc. The Hill Development Corporation will make suitable plans for scientific mining and provide appropriate finances and will be the marketing agency. After payment to Hill Development Corporation for its services and capital investment and to Mineral Development Corporation for services and to the mine owners the dividends, the net profits will be ploughed back to the mine workers. The mine owners pay about Rs. 10-15 per tonne for raising, breaking etc. only, but the sale price is Rs. 40-90 per tonne. Large portion of the difference after meeting transport, and royalty expenses goes to the mine owners, thus inadequate return goes to the persons actually carrying out mining.

Mining by modern and semi-mechanised methods will attract local educated persons especially when the return will be handsome.

Limestone of Mussoorie is high grade and should have controlled production and should be used for chemical and steel industry rather than paper and sugar industry where lower grades of limestone can be used. Diversification of lime industry will fetch much larger profits. Such units will be set up as subsidiary of the Hill Development Corporation.

In the interior e. g, at Bageshwar, and in Pindar and Ganga Valleys deposits of talc will be exploited for setting up mineral based industries at a number of centres. The earnings of such sub-corporations would be substantial in comparison to selling only raw material or semi-finished products.

Mining of magnesite will be taken up on the same pattern as limestone and will be marketed, in the first instance, to the immediate consumers.

Dairy

There are vast pasture lands at high altitudes which are capable of being harnessed for development of dairy, sheep breeding, mutton production etc. The system of dairy will also be village based. The Corporation will advance money for purchase of better breed of cows and buffallows instead of opening its own dairy and develop milk collection centres together with pasturisation facilities. The produce after meeting the local demand shall be supplied to the neighbouring areas. Appropriate system of cold storage vans shall be put on the road for long distance transport.

Tool and Spare Parts Manufacture

Numerous polytechnics have been opened or are being opened throughout the hills. The young men trained in these institutes can be given chance to manufacture tools and spares for various types of machinery which can sustain the transport impact. Such industries will be set up in most of the villages on the pattern of Haryana and Punjab.

There is vast scope of development of chemical and electronic component industries for which climatic factor is an important criterion.

The Hill Development Corporation has started working on sub-corporation basis viz. Teletronix at Bhimtal and Transcables at Haldwani. However, the control of Hill Development Corporation is not clearly defined. Instead of the member of the Hill Development Corporation being a member of the Board of Directors, he should have been made the Chairman for effective control.

Wool and Handloom Industries

The wool industry in the hill areas is very old and what is needed now is to increase the production on modern lines of refined goods, new designs, better dyings and better quality of wool. Training shall be imparted to all those who wish to undertake manufacture either collectively or individually. Weaving etc. shall be on cottage industry scale where the people can utilise their leisure time, thereby earning fair amount of money.

With stoppage of trade with Tibet, which was the main source of raw wool in the hill areas, the wool industry has got a serious setback. Arrangement will have to be made by the Hill Development Corporation for making raw wool or yarn available for supplementing the local production by organising trade with outside sources.

Since power is now fairly abundant in the hills introduction of Powerlooms would receive consideration. Improvement of designs and standardisation of quality will have to be given careful attention.

Tourism

Tourism shall be taken up as an industry rather than making peacemeal efforts. The present activities of the Hill Development Corporation which mainly comprise of organising package tours will continue and also expand. Twelve tourist buses will be on road by next season. Further development of tourism will be done on the following lines:

(i) Areas will be selected where tourists could come all the year round such as the Lake district of Uttar Pradesh Hills, i.e. district Nainital and other areas of easy approach where climate is ideal for establishing a flourising tourist industry. Added attractions for the pleasure of the tourists will not be overlooked. Accommodation will be constructed and furnished by the Hill Development Corporation and will be run as cooperative venture instead of engaging paid employees. Young students and others will run these establishments inclusive of catering. The local people will be imparted catering training and the profits will go to those who will manage the motels or tourists villas.

(ii) Package tours to places of religious and scenic interest are great attractions. Catering of good and wholesome food, soft drinks and accommodation facilities is a must and it should be developed as corporate venture.

(iii) Mountaineering tours shall be arranged on the pattern of Switzerland, Nepal etc.

(iv) Souvenir industry needs to be developed in the tourist areas so that the local art and craft get encouraged. The Hill Development Corporation will promote such units on corporate venture basis.

(v) The road transport will be gradually taken over by setting up of corporate units with appropriate coordination.

The Department of Tourism shall be responsible for promotional functions and provide technical and financial assistance.

Medicinal Herbs and Essential Oils

Medicinal herbs and perfume plant cultivation and manufacture of essential oils and extraction of active principles is yet another profitable line of business which can fetch quick returns and will be organised at suitable places as corporate ventures.

Processing of Agricultural and Horticultural Products

Processing of agricultural and horticultural products have special significance in hill areas of difficult terrain with poor means of transport. For proper return to the grower it is necessary to process the poor varieties of fruits in the form of jams, jallies, chutney etc. and to preserve better quality fruits for the use later. A proper survey of requirements and supplies in this respect is necessary. The Hill Development Corporation would undertake such a survey and establish viable units at selected centres.

Funds

One can judge that there are vast natural resources and possibilities of their utilisation are enormous. Fund requirements will, naturally, be of a sizable magnitude. The total requirement of funds would be known only after master plans have been prepared. With two crores share capital of the Hill Development Corporation, it would not be possible to embark the ambitious net work of industries and develop marketing and tourism. Funds will have to be found out by raising equity and from other sources.

Collaboration with Indian and Foreign Technicians

The Corporation will also look for technical and financial collaboration with Government or private agencies or advanced countries for the manufacture of more sophisticated instruments, watches, electronic components etc. with export bias.

10

RURAL SETTLEMENT AND HOUSE TYPES OF JAKHOLI BLOCK, GARHWAL HIMALAYA

M.S. Negi and B. P. Naithani

Settlement geography is concerned with building grouped around the permanent from dwelling, as well as the temporary camp at the hunter or harder.

There is enormous types of rural settlements in Garhwal region. The population size and settlement types vary from one basin to another. Present rural settlements of the Garhwal are clearly the product of a process of development and changing associated in changed social, economical and political conditions (Nand and Kumar, 1989). Types of permanent settlements vary from compact to clustered—farmstead or hamlated and scattered types representing the modes of human adjustment in the mountainous environment.

The Jakholi block, the rural settlements reflects its historical, cultural and socio-economic development of the region. The temple on the top of hills and in the middle site of the hillocks are the religious and spiritual sanctity of the dwellers.

The name of every region and place reflects its genetic characteristics. Jakholi, is the name of block headquarter which reflects in local language its meaning as where groups of hamlets (Jakh = where, Kholi= hamlet).

The Jakholi block is having the characters of the typical house types and types of rural settlement in all over the Garhwal Himalaya. The basic purpose of the study is highlight the rural settlements types, pattern, and their geomorphological sites in the Jakholi block. Secondly, to assess the typical house types and their variation with in altitudinal variations (1000-3700 m).

The study is based on the empirical methods of study. The 1:50000 topographical sheets are used for the analysis of rural settlements. The field study traverses, personal observations, revenue records and personal inquiries are the means for the study of typical house types. The field sketches and photographs are also helpful for carrying out the study.

Location

Jakholi block is located in the eastern part of Tehri district covering the area of 444.38 km^2. Geographically the block is bounded by long. 70°40' to79° 03′ E and Lat. 30° 18' to 30° 38° N. The headquarter is situated on the eastern ridge of Vishwan at the height of 2000 m.

Settlement History

Before independence this tract was being inhabited by migrated people from the south-western part of India, in the small group of houses on the favourite sites. Thus these villages had been organised into Ghar and ruled by Gharpati or Sardar.

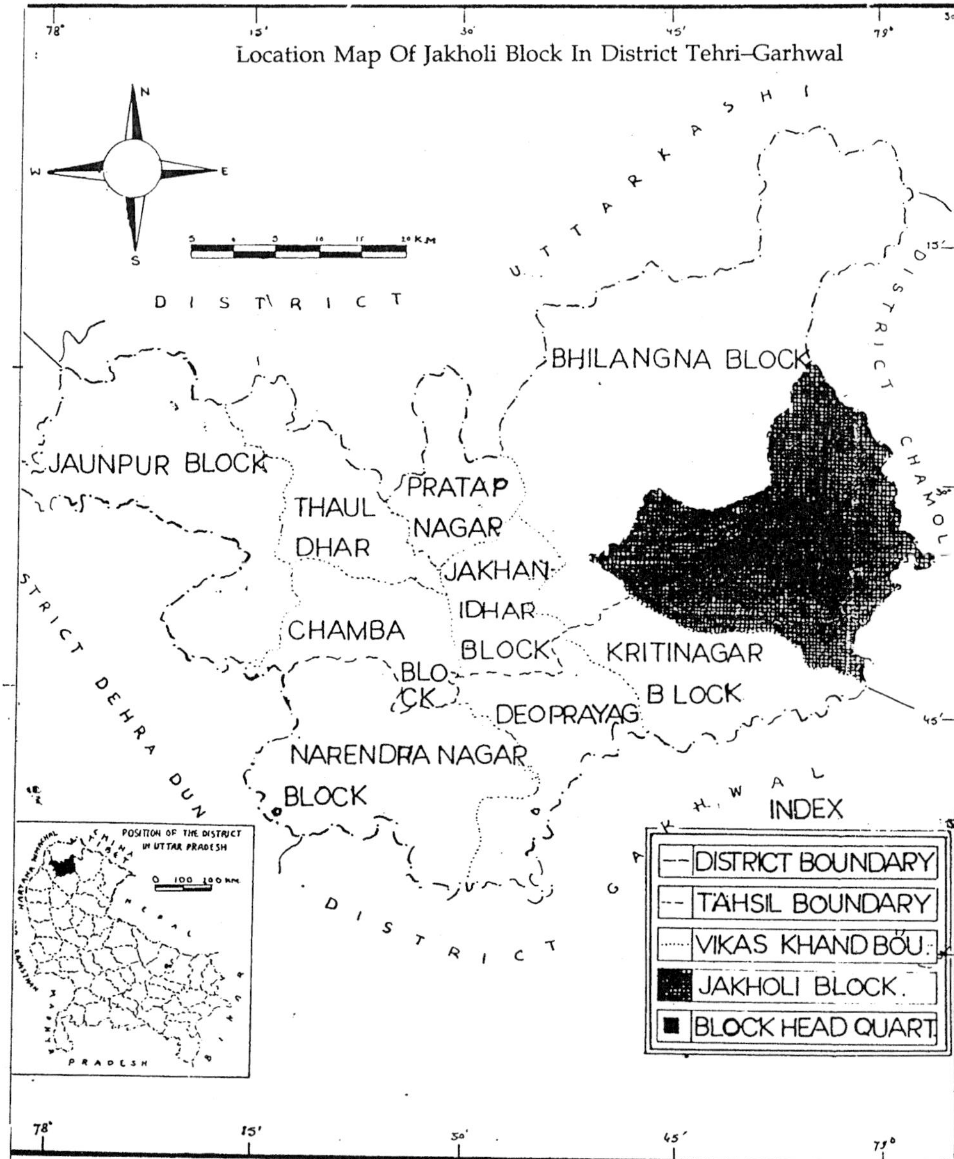
Location Map Of Jakholi Block In District Tehri–Garhwal
N
S
W
E
UTTARKASHI DISTRICT
BHILANGNA BLOCK
JAUNPUR BLOCK
THAUL DHAR
PRATAP NAGAR
JAKHANIDHAR BLOCK
CHAMBA BLOCK
KRITINAGAR BLOCK
DEOPRAYAG
NARENDRA NAGAR BLOCK
DISTRICT CHAMOLI
DISTRICT DEHRA DUN
GARHWAL DISTRICT
POSITION OF THE DISTRICT IN UTTAR PRADESH
PRADESH
INDEX
DISTRICT BOUNDARY
TAHSIL BOUNDARY
VIKAS KHAND BOU.
JAKHOLI BLOCK.
BLOCK HEAD QUART
78°
15'
30'
45'
79°

Before the inhabitation on the mountainous topography the area covered by dense forests and in between these, there were few scattered grass huts or thatches of tribes along the valleys. The migrated people i.e. Rajputs and Brahmins settled their settlements on the suitable sites according to their cultural social and racial systems. Later on these villages were known by their caste names such as Dakhwan Gaon, Dangwal gaon, Pujargaon, Saklana, Mehargaon, Ghanat gaon, Baman gaon, Kothiyara, Nawan gaon, and Anthwal gaon. At present, in these villages there is a social group of two to five caste people. They have constructed their settements on the good sites according to customs and ancient house plan. Most of the rural settlements were located nearby the water feature i.e. springs, tals, rivulets etc.

During the period of Sudarsan Shah and Manvendra Shah (1823 to 1944) the scheme of Bandovasta (revenue scheme of land measurement) had been started in which they had enrolled the account of Morisdar (landowner), Sirtha, Adhyaseem and also had determined the revenue tax (locally known as mamala). After 1945 the ruler have settled some new villages and forest banglow in the forest sites.

The existing settlements of the Jakholi block had been completed traditions and physical nature of the region. Presents villages are having well defined boundaries and settled on suitable sites from narrow valleys to high Himalayan ranges in the scattered manner throughout the block.

Distribution of Settlements

The distribution of rural settlements is based on the many sets of factors. The over all pattern of rural settlements in Jakholi block is guided by the drainage pattern, water divides and ridges, altitudes, slopes, water features and quality of land and soil. Besides this the cultural and social factors are also responsible for determining the pattern and distribution of settlements. Settlements are located almost exclusively along river valley whereever is cultivated land water is available.

The general distribution pattern of the rural settlement are followings :-

1. Higher density zone
2. Moderate density zone
3. Low density zone

1. Higher Density Zone: The higher density zone of settlement is along the river valleys (terraces) fertile patches of land, on spurs and irrigated areas. The southern part of the Helong valley, Luster Valley the bottoms of Nailchami Gad and the right bank of the Mandakini Valley from Sumari (Tilwara) to Rudraprayag. The density of the rural settlements is higher in per km^2. At present the density of settlement is also higher near the service center (Mayali, Chirbatiya, Phateru and Hoolanakhal) and administrative center (Jakholi and Ghansali).

2. Moderate Density Zone: The spurs and gentle mid-slopes of the hills are moderately settled. In this zone the agriculture and livestocks both are important occupations. All the settlements are permanent. The moderately density of zone is observed in eastern part of the Jakholi block, upper part of luster and Helong Gad.

3. Low Density Zone: The higher ridges of the Manakakhal, Dangi, Taila, Bhardar,

Hoolanakhal, Viswan, Chirbatiya Khal and Rai Khal are having low density in per Km^2. The main occupation of the zone is animal husbandry. The major factors for the responsible of low denstiy are dense forests, high altitude, cold climate, long growing period of crops and wild life. About 5% population of Jakholi block is lived in these zones.

Settlement Sites

The types of rural settlements in the Garhwal region are situated in different terrain, sunny aspects, springs, moderate slopes etc. which influence the types and pattern. The following table gives the general picture of settlement sites in the Jakholi block.

TABLE

S. No.	*Settlement Site*	*Aspect*	*Name of the valley/Ridge*	*Dominent villages*	*Remarks*
1.	River terrace	Valley	1. Luster Gad	Panyana, Rahar, Bhang, Tuneta, Thapala, Banoli, and Sumari	
			2. Heloun Gad	Koti, Mahargaon, Dhanyon, Makhat, Gharara and Mamni	
			3. Nailchani Gad	Mayalgaon, Theta Nailchami, Bhatwari, Tharali, Jakhnyali and Bahera	
			4. Changi Gad	Changi Badyarcura, Katura, Bodiyar, Pangriyana and Dobati	
			5. Dung Gad	Kot, Ghanotgaon, Futgari, Jakh and Thayali	
			6. Dangi Gad	Dangi, Uchhola and Syoor	
			7. Mandakini	Ratanpur, Medanpur and Pingari	
2.	River confluence	-	Luster and Helaon	Thapler	
3.	Khal or Saddle	Ridge	1. Bhardar ridge	Chirbatiya Khal and Hulankhal	
			2. Jammukhal	Jakholi and Sidhasore	
			3. Other	Quilakhal and Sonrakhal	
4.	Super		1. Luster Gad	Chopra, Kunyali, Pounthi, Khaliyan, Kapaniya, Chandi and Koliyara	
			2. Helaon Gad	Dewal, Gharara, Longa, Khariyal, Mayali and Bamngaon	
			3. Nailchami Gad	Bhatwara, Badiyargaon, Purala and Jhakhaniyali	
			4. Changi Gad	Khaseli, Kot, Pakh, and Jakh	
5.	Ridge	-	1. Between Helaon and Luster Gad	Thynkhar, Budna, Ludhiyar, Dhankurali, Uroli and Saklana	
			2. Between Helaon and Changi Gad	Hadiyana Malla, Sarpoli, Ganwari and Bhorgaon	
			3. Other	Sirwari, Khaliyan, Pulan, Jakhawari and Munyaghar	
6.	(A) Sunny aspect locally known as Taila	Southern	-	Taila, Nag Dharkot, Jakholi, Tailli Tharti, Badiyargaon, Kurcholla, Chopta, Nandwangaon, and Akhori	
	(B) Wet aspect locally known as Silla	Northern	-	Kapaniya, Tharti Silli, Rahar, Luthiyar and Akling	
7.	Religious centre	-	-	Nag Dharkot, Dewal, Pandav, Thalli, Pujargaon, Srikot and Hariyali	
8.	Tal (Lake)	-	Badhani Tal	Badhani Village	

Types of Settlements

Types and pattern of settlements are the most obvious expression of the synthesis of physical and cultural elements of a locality. The physical and cultural landscape of the Jakholi block is characterised by such a confusing variety as to be reflected in a distribution of highly complex settlement types.

On the basis of occupance, the rural settlement of Jakholi block can be divided into broad types :-

1. Seasonal settlements
2. Permanent settlements

1. Seasonal Settlements: Apart from the permanent settlement in the valley or low altitude there are some settlements which are fully and partially engaged in livestock raising. They live in the seasonal settlements on the upper part of the ridges locally known as Danda or Payar. The villagers used to go up with their cattles during the summer season (May to Sept.) of the year for seasonal occupation. They grow their potato with animal grazing. Generally Chhans are dispersed and single-storeyed huts along the ridges.

2. Permanent Settlements: The permanent habitations are occupied throughout the year. They are compact and cluster in spatial arrangement with double storyed houses. The lower storey is occupied by the cattle and goats while the upper storey is utilized for bedroom and keeping fuel and fodder. A small verandah in front of the bedroom is utilized for basking in the sunshine. Physical factors are more important for the development of types and patterns of settlement in the hilly terrain.

Pattern of Rural Settlement

Permanent habitations occupied throughout the year and seasonal habitation on the higher ridges can be classified into four types:

1. Compact settlements
2. Clustered settlements
3. Hamlated settlements
4. Isolated settlements

1. Compact Settlements: The villages, which are located on the spurs, river terraces are tighly compact because of the lack of space on the gentle slopy terrain. The inhabitants are avoid to construct their settlements on the good cultivated land. The settlements are located compactively on the spur where the least cultivated land is affected. The terrace settlements are also compactively settled at the break of slope points. Mostly concave slopes are most suitable. In some villages caste structure is also responsible factor for compactness. Main villages of this types are Bheti, Khaseti, Saruna, Quilakhal, Longa, Chonra, Bachwar Laluri, Mamni and Makhet.

2. Clustered Settlements: The compact settlement with a few groups of scattered houses is the most common type in big villages. The population size of these villages is between 500 and 1500 in the Jakholi block. Mostly mid-slope (concave and convex site) and on higher slopes at an altitude of about 1800 m are the prominent zones of clustered settlements. Between the altitude of 1200 and

1500 m the main winter settlements are compact and the rainy season habitations comprise scattered. The settlements which are located at an altitude of 1800 m . The Cowshades locally known as Chhan. Physical and social factors have manifested themselves in this village to develop a typical cluster-hamlet-farmstead type of complex settlement (Nand and Kumar, 1989). This type of complex settlements is found in Akhori, Paunthi, Budana, Handiyana Malla, Pakha, Anthwalgaon, Badiyar, Khaliyan, Jakholi, Arkund, Panana, Sumari Koti and Dhanyun.

3. Hamlated Settlements: Hamlated settlements are observed on the highly dissected and uneven slopy land where the cultivated area is pragmented. The cultivated area is found in pragmented pieces on gentle slopy terrain. These houses are constructed along a contour line. Usually there are rows of houses all joined together by common dividing partition walls. The grouping of settlement is restricted to related caste group. Here peasants perform intensive cultivation, hence they prefer to live close to their cultivated site. The example of these villages are Quarala, Jandarwala, Mahargaon Talla, Shyalduri, Sirani, Bachiwar, Koidhumka, Mathyagaon, Chopta etc.

4. Isolated Settlements: Isolated or scattered settlements are found in the ridges of Chirbaliya Kal, Viswan ridge, Hoolana, Bhardar and Rai Khal. The basic causes of the development of scattered villages are partition of families and lack of cultivated land in compact villages. The limited land resources are not adequate for food supply.Therefore, some members of the family became engaged in animal husbandry as a subsistence occupation. In the latter stages the families became settled near the small cultivated site or near the forested tracts where constant vigilance of fields is necessary to protect their crops from wild animals. The seasonal huts on the upper reaches are in this type. The main villages are Barsari, Syarasha, Bagiwala, Ghansara, Pangroli, Luthiyag, Badiarkura, Satole, Bhanser, Dobha and Gari.

House Types of Jakholi Block

The environment of the Garhwal Himalaya is most heterogence. House types reflect the influence of topography, climate, culture, economy and technology. As already stated the main occupations of the region are agriculture and animal husbandry which are distinct the local house types. The time factors is most important and responsible for changing pattern of rural house type. The house type of Jakholi Block has been divided into two parts. High altitude house types and valley house types. This classification is based on the altitudinal variations, climate and economic pattern. House types of high altitude are mostly seasonal and livestock raising. In the valley houses are double storey and Puccaka houses made with stone, slaty roof and wood. The close relationship between house type and landscape in Jakholi block reflects the harmony of man and nature. The Garhwalis construct their houses in relation to hazard, wind, sun, shade and as well as topography (Nand and Kumar, 1989).

On the basis of shape, layout, house material and architect the house types of Jakholi block may be classified into following three types:

1. Single storey houses
2. Double storey houses
3. Tripple storey houses

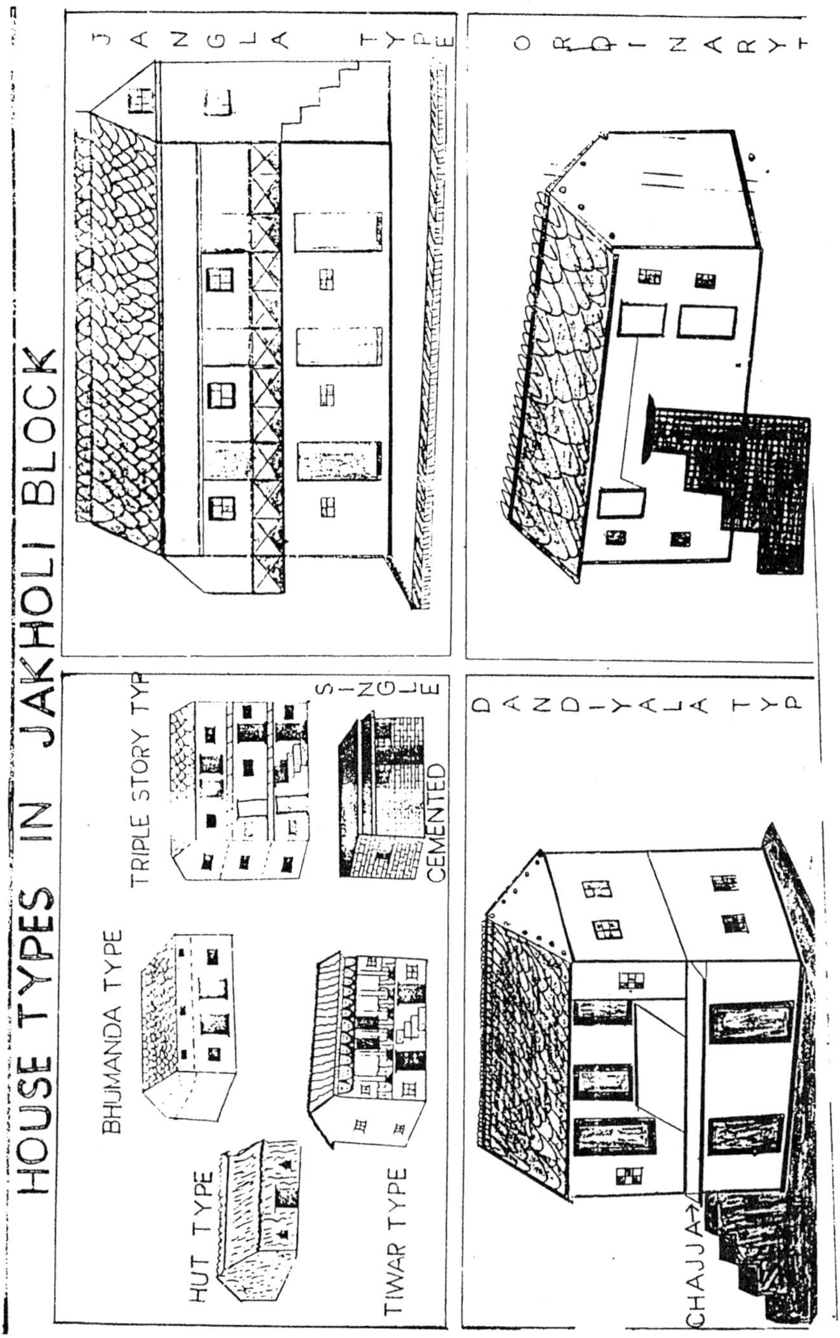
HOUSE TYPES IN JAKHOLI BLOCK
JANGLA TYPE
ORDINARY TYPE
HUT TYPE
BHUMANDA TYPE
TRIPLE STORY TYP
TIWAR TYPE
SINGLE CEMENTED
DANDIYALA TYP
CHAJJA

1. Single Storey Houses: Most of the single storey houses are cattle shades and which are spread over the upper reaches of mountain ridges. In Jakholi block there are three types of single storey houses and cattle shades: (i) Bhumunda type (ii) Shal (cattle shade) and (iii) Hut type. The single storey cattle shades locally known as Murida or Chani. Most of the Murda are located on the Bhardar range and Chirbatiyar Khal range. The houses are built of stone, wood and grasses. The walls are made of stones (granite, gneisse and schist) while the roofs are made by wood with grass cover. The floor of the Chani are made of slate or split stone flags. There are two rooms in single storey houses. The large size room is for cattles, goats and sheep. Besides this the small size room plastered with mud and cowdung is for people.

2. Double Storey Houses: The double storey houses are very common in the Jakholi block. They have two rooms, one long room is in the ground floor and two are on the first floor. The ground floor generally is used for cattle shade, fuel and fodder. One room on the first floor is utilized as bedroom and the other as kitchen and store. On the basis of shape and size the double storey houses are of Jakholi block are classified into following types :-

1. Dandiyala type
2. Tibari type
3. Nimdri type
4. Jangla type
5. Ordinary type

In Dandiyala houses have an open verandah called dandiyala which is supported by wooden posts in front of the rooms on the first floor. When this varandah is ornamented with four pillars of carved wood, one being at either end and two trisecting the space between them, it is called Tibari.

3. Tripple Storey Houses: The triple storey houses are generally found on the fertile tibur terraces and near the block head-quarters, road and service centre. These houses are locally known as Pacca Ghar which are made by rich people. Thatched and slate roofs replaced by tin and cement roofs. The examples of these houses are found in the villages of Purwal gaon, Kot, Dang, Kharseli, Cheli and Munyaghar.

11

AGRICULTURAL DEVELOPMENT IN A HILL REGION: AN APPRAISAL

K. N. Joshi

The generation and sustenance of economic growth, especially in the early stages of most of the developing countries are, to a large extent, determined by the performance of the agricultural sector[1]. An agricultural sector, non-responsive to stimuli, or unstable in nature, may substantially impede steady growth of the economy. From this point of view, an appraisal of agricultural development in hill areas, particularly in North-Western Garhwal Himalayan Region is, therefore, most necessary. Moreover, it is an essential aspect of growth studies.

Situated in the North-Western Himalaya's in U.P., the economy of Garhwal region is predominantly agricultural. The share of agricultural sector in the total output of Pauri +Chamoli, Tehri +Uttarkashi and Dehradun districts at current prices in 1975-76 was 72.8, 68.8 and 59.7 per cent, respectively[2]. An overwhelming majority of the population of the region is rural and 69.3 per cent of its work force is directly engaged in agriculture; but agriculture in the region is still in its primitive stage or of mere subsistance type. Relief, topography, soils, climate and social structure are some important factors not best suited for the growth of agriculture in the region. The development and growth of this sector in the hills in the past had not been successful in keeping pace with the development in the other parts of the country/ state, and even with the requirements of the region itself. The low productivity in agriculture and foodgrain deficiency ranging from 24 to 32 per cent[3] annually are resulting an over-dependence of the region upon imports of foodgrains. Insufficient production of milk and milk products, fruits and vegetables further increases this deficit. As a result the hill region reveals a true picture of unfavourable terms of trade. The present sluggish growth of agricultural economy resulted poverty, inequality, unemployment and mass resentment in the area. Therefore, some immediate strategic modifications in the present planning system are considered necessary.

In this light the present paper, aims to highlight the recent trends in agricultural development in the hills and to suggest alternative strategy for proper development and growth of the economy in general, and agriculture in particular, in the area.

Data-base and Methodology

The study is based on secondary data collected from various types of statistical publications at state and district level. The analysis is centered on principal crops only, which cover more than 90 per cent of the cropped area of the region. For eliminating short period fluctuations, three year moving averages have been derived, and then the compound growth rates of important variables have been computed by fitting exponential functions of the type:

$$P_t = P_o R^t$$

Where

P_t = Production, Productivity of each crop after time t.

P_o = Initial value of production, productivity at time t=0 (reference time).

$R = 1 + \frac{r}{100}$; and

r = Compound growth rate, given by

r = (antilog b -1) x 100

where,

$b= \frac{1}{t}$ Pt/Po).

TABLE 1: **Percentage Distribution of Number and Area of Operational Holdings in Garhwal Region and U.P.**

Size class in Hectares	*Number in 1976-77*		*Operated area in 1976-77*	
	Garhwal Region	*U.P*	*Garhwal Reigon*	*U.P*
Marginal Holdings less than 1 hect.	64.7	69.4	21.0	23.9
Small holdings 1 to 2 hect.	19.8	16.4	26.0	21.6
Semi-medium holdings 2 to 4 hect.	11.8	9.6	29.5	24.8
Medium holdings 2 to 4 hect.	3.5	4.1	18.1	22.3
Large holdings above 10 hect.	0.2	0.5	5.4	7.4
Total	100.00	100.00	100.00	100.00

Source : Agricultural Census in U.P. , 1976-77, Pt I & II Board of Revenue, U.P. Lucknow.

TABLE 2: **Average Size of Holding in Garhwal Region (area in hectares)**

Districts	*1970-71*	*1976-77*	*1980-81*
Pauri	0.90	1.30	1.33
Tehri	0.89	0.91	0.82
Chamoli	0.64	0.94	0.90
Uttarkashi	1.13	0.89	0.92
Dehradun	1.13	1.11	0.98
Garhwal Region	0.91	106	1.01
U.P.	1.16	1.05	1.01

Sources : 1. Agricultural Census in U.P. 1970-71 & 1976-77.
2. Agricultural Situation in India, June 1984.

PART-I

(i) Man-Land Ratio : The man-land ratio in the hills in general and Garhwal regioı in particular, is extremely poor, and the land is relatively less fertile. Of the tota reported area of 3303.8 thousand hectare, only 10 per cent is under cultivation ranging from 4.1 per cent inUttarkashi to 17.9 per cent in Dehradun, as against 57.9 per cent in the State and 46.7 per cent in the country[5]. Total net sown area in Garhwal region in 1982-83 was 311.5 thousand hectares against 17266 thousand hectares in U.P.,[6] dividing it by the total population of the region and the state, the percapita N.S.A. comes to 0.13 hectare and 0.15 hectare for the region and the State respectively. If calculated on the basis of the population wholly dependant on agriculture, it comes to 0.48 hectare in the region as against 0.72 hectares in the State.[7]

(ii) Agricultural Holdings : The agrarian structure of rural Garhwal is dominated by marginal and small farms which are fragmented, and more than 2/3 of them are scattered. Table 1 show that 84.5 per cent of the holdings covering about 47 per cent of the operated area are marginal and small in size. The number of medium and large holdings is small (15.5%)though they cover comparatively large cultivated area (53%). The average size of holdings is also very small. Though it increased from 0.91 to 1.01 hectares during 1971-81, but the change did not occur uniformly in each district of the region Table 2. It increased in Pauri and Chamoli but decreased in Tehri, Uttarkashi and Dehradun districts. Average size of holding was equal in both the region and the state in 1981, but the inferior quality of land is a serious handicap here.

(iii) Cropping Pattern : The main characteristics of the cropping pattern in the hills is that the major portion of the gross cropped area remains under foodgrains, but pulses, oilseeds, and other commercial crops have not been given due importance.

Table 3 shows that out of the total cropped area of 481 thousand hectares in the region in 1982-83, 443 thousand hectares (92 per cent) was accounted by cereals, and only 38 thousand hectares (8 per cent) was under non-food or commercial crops, whereas in the state foodgrain and commercial crops accounted 80 per cent and 20 per cent respectively. It is also important to note that from 1971-72 to 1982-83 the area under non-food crops in the region increased only by 3 percentage points, which is not significant to lead the agriculture on the progressive path. In brief, the cropping pattern in hill areas has not undergone any substantial change showing significant diversification.

TABLE 3: **District-wist Percentage Distribution of Cropped Area Among Different Crops in Garhwal Region (1982-83)**

Crops	*Pauri*	*Tehri*	*Chamoli*	*Uttarkashi*	*Dehradun*	*Garhwal*	*U.P.*
1. Paddy	18.5	14.4	26.4	23.2	18.3	19.3	20.5
2. Mandua	25.8	19.0	24.0	17.5	4.7	19.3	0.6
3. Sanwa	15.4	20.3	5.4	8.0	5.0	12.3	0.5
4. Maize	1.6	1.4	0.4	0.9	14.4	3.6	4.4
5. Wheat	29.7	33.4	32.3	30.2	32.3	31.5	33.6
6. Barley	6.0	4.3	4.9	1.7	2.6	4.3	2.5
7. Other Cereals	—	—	—	—	—	—	6.1
8. Pulses	0.8	1.6	0.3	1.4	4.6	1.7	12.0
9. Nonfood Crops	2.2	5.6	6.2	16.8	18.1	8.0	19.8
Total	100.00	100.00	100.00	100.00	100.00	100.00	100.00

Source: Uttar Pradesh Ke Krishi Ankare, 1982-83.

(iv) Production of Foodgrains: Although the economy of Garhwal region is basically agricultural, but the region is still a food deficit zone. The supply of foodgrains produced locally is not sufficient for meeting the existing requirements of the people. Planned efforts during past decades, though, succeeded in increasing the agricultural production but not so significantly as in other parts of the state and country. It is evident by the fact that during 1971-74 to 1980-83 foodgrain production in Garhwal region increased at a compound rate of 0.79 per cent as against 3.94 per cent in the state. Table 4 shows that production of paddy and maize have shown only a substantial growth of 1.65 and 3.93 per cent annually, while the production of other important foodcrops of the region, such as mandua and wheat, have decreased at a compound rate of 3.06 and 0.34 per cent respectively. Production of Sanwa, though, increased at an annual rate of 0.24 per cent but it covers a short period, i.e. from 1976-79 to 1980-83. No doubt, production of cereals in the region increased at a rate of 0.86 per cent, but the production of pulses decreased sharply by a rate of 4.72 per cent . As a result, the rate of growth in respect to total foodgrain production lowered. It is important to note that growth rates of production for most of the crops in the region are lower than the state. The district-wise study also reflects a picture of asymmetrical growth in production of different crops. Out of five districts only three have recorded an increase in foodgrain production, though substantially by Uttarkashi and Dehradun districts only. These two districts recorded an annual growth of 2.39 and 4.44 per cent, respectively. Tehri, on the other hand, showed only marginal increase in its foodgrain production, particularly, due to inclusion of sanwa production in the current year, which was absent in the base year figures. Contrary to this, food grain production decreased in Pauri and Chamoli districts at a compund rate of 0.17 and 1.01 percent, respectively, during the same period.

TABLE 4: **Compound Growth Rates of Production of Major Crops in Garhwal Region From 1971-74 to 1980-83**

Districts	*Paddy*	*Mandua*	*Maize*	*Sanwa**	*Wheat*	*Barley*	*Total Cereals*	*Pulses*	*Total Foodgrains*
Pauri	0.02	—2.92	—0.57	1.67	—2.67	0.95	—0.14	—4.23	0.17
Tehri	—0.77	—6.11	—0.20	—0.83	—0.17	—2.42	—1.50	—6.77	0.41
Chamoli	2.91	—3.30	—11.25	0.93	—2.19	—6.90	—0.98	—19.16	—1.01
Uttarkashi	16.68	1.50	—0.73	—1.65	0.21	—8.54	2.50	—6.23	2.39
Dehradun	3.61	2.70	6.23	3.45	5.09	3.59	4.71	—3.53	4.44
Garhwal Region	1.65	—3.06	3.93	0.24	—0.34	—2.30	0.86	—4.72	0.79
U.P.	5.44	—4.43	—1.79	7.43	7.06	—3.87	4.57	—0.52	3.94

*Growth rates of Sanwa for Garhwal Region are from 1976-79 to 1980-83.

Data Source: 1. Parvatiya Shatron me Krishi Utpadan.
2. Ritu aur Upaz Report.
3. Uttar Pradesh Ke Krishi Ankare.

(v) Productivity of Major Crops: The changes in the productivity of crops in the region and the state overtime have been shown in Table 5. Productivity of paddy only increased in all the districts of the region as well as in the State, the compound growth rates were being 2.23 and 3.07 per cent, respectively, for the region and the State. Productivity of wheat increased in all the segments, except Chamoli district, but the growth was not attractive. It increased in the region at a rate of 1.16 per cent, as against 4.08 per cent in the State. Thus, the growth in productivity of paddy and wheat could not keep up with the State's average growth. Maize recorded the highest growth in productivity in the region, while barley the lowest. This ran counter to the pattern of crop productivities in the state. Mandua recorded a decrease in productivity but less than the state. Taking all the crops together, the productivity in the region increased at a rate of 1.56 per cent as against 3.51 per cent in state. The district-wise analysis of productivity also reveals wide differences. Productivity of almost all the crops in Uttarkashi and Dehradun districts increased substantially, but it could not increase so smoothly in other districts of the region.

TABLE 5: **Compound Growth Rates of Productivity of Major Corps in Garhwal Region From 1971-74 to 1980-83**

Districts	*Paddy*	*Madua*	*Maize*	*Sanwa**	*Wheat*	*Barley*	*Pulses*
Pauri	1.52	—1.35	—1.33	—0.49	0.51	—0.15	—0.84
Tehri	2.85	—2.13	—0.78	—1.54	1.21	1.77	3.42
Chamoli	2.31	—0.92	—0.88	—4.98	—0.16	—0.48	0.61
Uttarkashi	3.35	2.68	0.75	2.85	0.74	1.34	3.13
Dehradun	1.29	4.54	4.65	—4.85	3.05	5.94	0.23
Garhwal Region	2.23	—0.45	3.11	—1.77	1.16	0.30	1.56
U.P	3.07	—0.56	0.48	5.90	4.08	2.33	3.51

*Growth rates of Sanwa for Garhwal Region are from 1976-79 to 1980-83.

Data Source: 1. Parvatiya Shatron me Krishi Utpadan.
2. Ritu aur Upaz Report.
3. Uttar Pradesh Ke Krishi Ankare.

PART II

(vi) Growth in Irrigation Facilities: Because of the positive correlation between provision of irrigation and production of crops development of irrigation has been the most crucial part of agriculture development programmes for achieving more production and self-reliance. Though it is difficult to construct various means of irrigation in the hills owing to its geographical conditions, yet, a lot of efforts have been made during plan period to construct canals, guls, reserviors, pumpsets and tubewells in the public and private sector throughout the region. However, tubewells are confined mainly to the plains of Dehradun and Pauri districts. Due to lack of electrification in rural areas, water lifting devices, which are now considered most suitable for hilly areas, are also not been found every where. Of course, the past development of means of irrigation in the region provided water to comparatively larger portion of the net sown area, but as Table 6 shows, only 16.21 per cent of the NSA was only irrigated in Garhwal region in 1980-83 as against 55.83 per cent in the State. A look at the compound growth rates of net irrigated area and gross irrigated area also reveal that their growth in the Garhwal region are also less than their region growth in the State. From 1971-74 to 1980-83 NIA and GIA in the region increased at a rate of 2.0 and 2.2 per cent annually as against 2.9 and

4.1 per cent in the State. As regard their growth in the constituent districts it is observed that Uttarkashi and Dehradun recorded higher growth in NIA than its growth in the region and the State, but the performance of the other districts was very poor. In Pauri and Chamoli districts the growth rates are found almost negligible. In the case of gross irrigated area the segments once again repeated the same trend of growth. As for the growth of irrigated area under chief crops (paddy and wheat) it is found that irrigated area under paddy and wheat have not increased so fast as it has increased in the State, e.g. irrigated area under paddy and wheat increased in the region at a compound rate of 2.0 and 2.45 per cent as against 5.84 and 4.83 per cent in the State. It is important to note that, in this case too, Uttarkashi and Dehradun districts were again in a better position than the other districts of the region.

TABLE 6: **Growth of Irrigation in Garhwal Region**

Districts	*NIA as% of NSA 1971-74*	*NIA as% of NSA 1980-83*	*CGR of NIA 1980-83 over 1971-74*	*CGR Of GIA 1980-83 over 1971-74*	*CGR Of IA under Paddy*	*CGR of IA under Wheat*	*CGR of IA under Maize*
Pauri	8.13	8.06	0.04	0.9	0.56	0.61	8.31
Tehri	13.15	15.23	1.1	1.1	0.97	2.41	0.57
Chamoli	5.0	5.44	0.01	0.9	1.75	0.05	1.55
Uttarkashi	14.65	16.08	3.2	4.7	5.25	4.31	—3.0
Dehradun	32.35	40.93	4.5	2.7	3.02	3.60	0.43
Garhwal Region	13.72	16.21	2.0	2.2	2.0	2.45	1.09
U.P.	41.63	55.83	2.9	4.1	5.84	4.83	2.70

(vii) Area Under High Yielding Varieties: For increasing the production of foodgrains rapidly in the region High Yielding Variety Programme has been given special importance since Fourth Five Year Plan. The State Govt. had also granted subsidy on the transportation of H.Y.V. seeds to the hill areas, so that these might be made available to the farmers at a cheaper rate. However, small size of land holdings & tiny terraces, lack of irrigation facilities, and poor infrastructure were proved to be the main limiting factors. As a result, these quality seeds could not be used so extensively throughout the region as in the plains of western U.P. As evident by the Table 7, only 20.3 percent of the gross cropped area in Garhwal region was under H.Y.V., as against 37.5 per cent in the State. Amongst the constituent districts of the region, the proportion of area under H.Y.V. was highest in Dehradun (37.1 per cent) and lowest in Pauri (10.3 per cent). As for the growth of area under H.Y.V. of different crops is concerned it is observed that the area under H.Y.V. of paddy, wheat and maize increased at a compound rate of 11.92, 10.38 and 11.03 per cent in the region. Constituent districts also recorded remarkable growth during 1971-74 to 1980-83. It was particularly because of the small base figures. It is also important to note that the growth of area under H.Y.V. of different crops is not found to be uniform in each district.

(viii) Consumption of Fertilizers: The H.Y.Vs are characterised by their responsiveness to higher doses of chemical fertilizers, which is a costly input. Though, the per hectare consumption of fertilizers, as shown in Table 7, is increasing in the region since 1971-72, but it is still too low in comparison to the average of the state and other hill regions of the country. In 1982-83 the region's per hectare consumption of fertilizers stood at 6 kg., ranging from 1.45 kg. in Tehri and 21.62 kg in Dehradun; whereas it was 58 kg (more than 9 times) in the state. As regards the growth in fertilizers consumption during 1971-74 to 1980-83, it is observed that it increased at a compound rate of 2.79 per cent in Garhwal region as against 10.35 percent in the State. Similarly, the per hactare consumption in the region increased at an annual rate of 3.49 per cent against 9.21 per cent in the State.

TABLE 7 : **Growth in the Area Under H.Y.V and Consumption of Fertilizers (1971-74 to 1980-83)**

Districts	*Area under H.Y.V. as % of TCA 1971-72*	*Area under H.Y.V. as % of TCA 1982-83*	*CGR of Area under H.Y.V. Paddy*	*CGR of Area under H.Y.V. Wheat*	*CGR of Area under H.Y.V. Maize*	*Per Hectare Consumption of Fertilizers*	
						1971-72	*1982-83*
Pauri	3.9	10.3	4.32	10.19	7.75	1.18	1.98
Tehri	3.4	22.1	15.23	9.23	16.89	1.10	1.45
Chamoli	2.4	17.5	10.71	13.74	3.91	1.40	3.09
Uttarkashi	6.9	21.6	14.97	8.36	11.95	2.70	6.45
Dehradun	7.1	37.1	13.54	11.17	11.45	17.80	21.62
Garhwal Region	4.3	20.3	11.92	10.38	11.03	4.05	5.99
U.P	26.6	37.5	NA	NA	NA	20.0	57.62

Source of base data: 1. Office Records of Deputy Director of Agriculture, Pauri.
2. Parvatiya Shetron me Krishi Utpadan.
3. Uttar Pradesh Ke Krishi Ankare.
4. Statistical Diary of Garhwal Mandal 1984-85.
5. Statistical Diary of U.P. 1984-85.

(ix) Agricultural Credit : As the non-institutional credit has proved to be more of a drag than help, the post-independence national policy in regard to agricultural credit has been to institutionalise it progressively by setting up new credit institutions and also by strengthening the co-operative credit societies in the region. Therefore, a large number of financial institutions have come into existence, expanded their operations and strengthened the credit structure in the region. As shown in Table 8, in 1973-74, 25 cooperative bank offices and 77 commercial bank offices were working in the region which rose to 78 and 207, respectively, in 1984-85 and 1982-83. Similarly, the loan distributed by cooperative societies increased from Rs. 26.675 lakh in 1973-74 to Rs. 108.901 lakh in 1984-85. Besides, loan distributed by Zila Sahkari Banks increased from Rs. 32.777 lakh in 1973-74 to Rs. 82.386 lakh in 1981-82. However, commercial banks sanctioned 12.4 per cent of their total loans to the agricultural sector in 1984. These changes in the credit structure reduced the dependence of farmers on private moneylenders and affected agricultural operations favourably.

TABLE 8: **Progress of Financial Institutions and Distribution of Loans in Garhwal Region.**

Districts	*1973-74*		*1984-85*	*1982-83*	*Loan Distribution by Primary Coop Societies in Lakh Rs.*		*Loan Distribution by Zila Sahkari Bank in Lakh Rs.*	
	No. of Zila Sahkari Bank	*No. of Commer-cial Bank*	*Zila Sahkari Bank*	*Commercial Bank*	*1973-74*	*1984-85*	*1973-74*	*1981-82*
Pauri	6	11	17	39	4.129	13.699	3.919	32.055
Tehri	8	9	19	32	10.195	16.571	7.030	24.952
Chamoli	3	4	17	28	2.479	18.939	2.156	20.422
Uttarkashi	3	3	11	16	2.773	10.682	3.041	16.669
Dehradun	57	50	14	92	13.201	22.495	10.529	14.803
Garhwal Region	25	77	78	207	32.777	82.386	26.675	108.901

Source: 1. Draft Fifth Five Year Plans of Pauri, Tehri, Chamoli, Uttarkashi and Dehradun districts.
2. Statistical Diary of Garhwal Mandal 1984 and 1985.
3. Statistical Diary of Hill Region, 1983.

PART III

The above analysis of agricultural development in Garhwal Himalayas leads us to the conclusion that agriculture in the hill region could not progress so appreciably as in the plains of U.P. It is, therefore necessary to comment briefly upon the development strategy pursued in the light of problems and prospects of agricultural development. What is all the more surprising is that even three decades experience of economic planning has yet not awakened the planners and the community in general and the leadership of the region in particular to the basic short-comings of the development strategy hitherto pursued and the changes which need to be affected to rid the region of poverty, squalor and disease and to exploit fully the immense resource potential of the region.

The region is rich enough in resources to maintain its population at reasonable standard. However, one finds the acutest pockets of poverty in this region. Much of the resources have been allowed to go waste for various reasons. The yield per hectare is very low. The proportion of area irrigated is the lowest in U.P. The use of H.Y.V. Fertilizer technology has made a beginning in the region but the achievement is not very encouraging. Hydro-power resources of the region have hardly been used for irrigation through suitable location specific micro-hydro-electric units. The human resources have remained mostly undeveloped despite a big effort for educational institutions of various kinds.

A closer study of development activities in the region and the various plan programmes and priorities reveals a sorry imitation of the plan priorities of the State and the country, irrespective of any economic assessment of their suitability to the hilly region and its needs. The plans and priorities and goals have been misconceived and lop-sided and no attempt have been made to adjust them to the area–specific problems of the region and search for their solution. A case by way of illustration may be cited here. There is population pressure on land in the region which in the present state cannot sustain the region's high working populations. As a result, over the decades the region has been denuded of its young, enterprizing and productive workers for want of suitable job opportunities. This has further affected adversely the agricultural operations, because women generally do the agricultural operations in addition to their over-burdened household duties, and old people cannot be expected to provide energetic initiative and efficient management. What could be done under ideal area specific programmes of action has been nobody's concern. If resource potential of each sub-region, say a cluster of villages or a block, is made the basis of micro-level planning of the sub-region embracing agricultural and allied activities including household industries, there is no reason why the employment opportunities cannot be expanded through progressive agricultural operations.

An alternative planning strategy can be devised by inducting rational changes in the pattern of subsistence, foodgrain dominated agriculture through a better land-use pattern, where there is an optimal mix of foodgrains, horticultural and off season vegetable crops depending on climatic conditions, and a host of agro-based small scale units processing the local produce into semi-manufactured, unsophisticated goods for use of local population and also suitably devised handicraft products to be exported outside the region. This is the only way in which economy of the hilly districts of Garhwal region can be diversified at the present stage of development.

REFERENCES

1. Rains, G. and Fei, H., "A Theory of Economic Development" The American Economic Review, Vol. LI, No. 4, Sept. 1961.
2. Sinha, R.C., "Some Observations on the Economic Identity and Development of the Hill Region of U.P." in Popota, T.S. (ed.) Development of Hill Areas, Issues and Approches, p. 159.
3. Juyal, B.N., "Inter-Regional Inequality and the Process of National Development : The Case of Uttarakhand" in Popota T.S. (ed.) Development of Hill Areas, Issues and Approaches , p. 50.
4. Directorate of Agriculture, U.P., Lucknow, Uttar Pradesh Ke Krishi Ankare, 1982-83.
5. Dasai, Basant, 'A Study of Rural Economics', Himalaya Publishing House, Bombay, p. 14.
6. Directorate of Agriculture, U.P., Lucknow, Uttar Pradesh Ke Krishi Ankare, 1982-83.
7. Joshi, K.N., 'Agricultural Development in Garhwal Region since 1971-72' Unpublished Thesis, Garhwal University, 1987, p. 124.

12

RURAL LANDUSE PATTERN OF CENTRAL GARHWAL HIMALAYA: A CASE STUDY OF SOUTH-EAST TEHRI DISTRICT

D.D. Maithani and R.K. Verma

Introduction

In the Himalayan region, the land resource is the bed rock of the economy because agriculture is the main occupation of the people. The pattern of land utilization is conditioned by two sets of factors:

The physical factors like topography, climate and soils which set the broad limits upon the capabilities of the land, and other is the human factors like length of occupation of the area, density of population, social and economic factor especially system of land tenure and the technological levels of the people.

The Study Area

The area of present study, South-East Tehri-Garhwal is in the central Garhwal Himalaya of Uttar Pradesh. Geographically the area is bounded by 30° 5' N. to 30° 37' N Latitudes and 78° 15'E to 79° 3'E longitudes, comprises one Tehsil and four blocks, via, Jakholi, Kirtinagar, Deoprayag and Narendranagar. This region has an area of 1789.92 km^2 in the form of irregular rectangle with a density of 115 persons per km^2.

Trends of Landuse Pattern

The present trend of land utilization in the region, to consider the numbers of changes that has taken place during the past 30 to 40 years. Surveys regarding these changes in landuse undertaken in some pattis[x] of South-East Tehri-Garhwal district indicate that there have occurred slight changes in the pattern of landuse which are indicative of the general trend in landuse. The net area sown has not increased appreciably except valley bottoms and gently sloping spurs. Double crops area has shown an increase of about 20% and total cropped area has shown some increase due to the rapidly increasing population of the region.

The irrigated area has increased at about 8% mainly, through the extension of 'guls' or canals. The area under current fallows showed some decrease, because of a sudden increase in maize, rice and Mandua areas. The percentage of cultivated area varies from Patti to Patti in the region.

---

x A Patti is a sub-division of a Pargana.

General Pattern of Landuse

An attempt has been made here to work out the area and percentages of various uses of land in 1988-89 and the distributional figures of land use pattern have been given in the Table 1. The broad classification of landuse is discussed below :

TABLE 1: **Landuse Categories**

Altitudinal Zone (M)	*Cultivated land km²*	*%*	*Waste land km²*	*%*	*Forest land km²*	*%*
Below 1000	33.43	23.39	9.87	9.58	2.33	0.39
1000-1500	58.75	41.09	41.75	40.55	15.19	20.06
1500-2000	45.92	32.13	33.25	32.29	24.23	33.72
2000-2500	3.85	2.69	15.97	15.52	28.53	38.68
Above 2500	1.01	0.70	2.13	2.06	5.42	7.15
Total	142.96	100.00	102.97	100.00	75.70	100.00

Crop Land

In the study region about 50.48% of total area is under cultivation. Such areas are mainly confined to the river terraces of Alaknanda, Mandakini, Bhagirathi, Bhilangana, Laster valleys and gently sloping spurs. The landuse map shows that Deoprayag sub-division[x] (Block) has maximum 29.97% net cultivated area, while Kirtinagar, Jakholi and Narendranagar sub-division have less area, 24.55%, 23.45% and 22.03% respectively due to large area under forests and pasture land.

Changes in cultivated land are observed from the upper valley to lower valley. These changes are due to the altitude, slope, formation of soil and water availability in the high altitude (above 1000m.) Locally, irrigated land is known as 'Sera' or 'Talaon' and 'Seragoan' and Unirrigated land is known as 'Ukhar' and 'Ukhari sari'. Maximum unirrrigated land lie in Kirtinagar and Deoprayag sub-division, accounting for 28.9% and 26.3% (Table 2) which is 3.5% and 2.8% of the district respectively.

TABLE 2: **Agricultural Land**

Sub-division	*Mixed land*		*Unirrigated land*	
	% of sub division	*% of district*	*% of sub division*	*% of district*
1. Jakholi	88.6	9.8	10.8	1.11
2. Kirtinagar	73.0	6.9	28.9	3.5
3. Deoprayag	75.3	11.2	26.3	2.8
4. Narendranagar	92.1	10.7	8.9	1.6

Source: Census Handbook of Tehri-Garhwal (B), 1981.

In the region, some agricultural land in the villages are mixed both irrigated and unirrigated. Such villages are maximum number in every sub-division possesses more than 70% (Table 2) of such as Kothiyara, Launga, Barsera, Tuneta, Sumari, Seragaon, Suppar and Maletha etc. Many smaller patches of cultivated land are observed in this area, while most of the land is under waste or scrub land. The

x Sub-division is a Block of district.

Northern part (above 1000 m.) of the W/E Bhardar Patti, Lasya Patti, Bangar Patti, Badiyargad Patti, Nailchami and Hindan Patti are having the lowest area of the cultivated land. The irrigated tracts in the region are situated only near the bank of streams and rivulets, which are known as 'Gadhera' and 'Gards'.

Waste Land

In the region, land which has been measured at any settlement and recorded as the property of individuals is called 'Nap land' or measured land, 'Benap' is unmeasured land. The most extensive waste lands includes rocky land, steep cliff slopes, pasture land, scrubland and area that comes under rivulets and rivers. About 20.19% (102.97 km^2) of land falls under this category of the region. Main areas are Agrakhal-Kunjapuri range, Ghandiyal-Lawa range, Chandrabadni range, Rajbonga, Ruddukhal- Panwali and Hulanakhal Range. Some small patches of waste land are marked in the lower valley bottoms. In remote places where the possibilities of cultivation are not present, as in large tracts of region, the surface of ground approximates its natural condition. Some are not available for cultivation occupied by villages, pattis under bed of streams, river and other water bodies and barren lands. The crop production is not satisfactory. The water divides are generally dry. Long patches of waste land are generally found along the fringe of reserved forests.

Forest Land

In the study region, about 31.39% (75.70 km^2) area is under forest, the main forest contain Deodar (Cederw), Chir (Pinus longifolia), Khair (Acacia catchu), Sal (Sheroa robusta), Seman (Bombax matabaricum) and Oak (quercusspp) etc. Maximum growth of Chir is below 2000 m. as Alakananda, Mandakani and Bhilangana reserve forests. Between 1800 m. to 2500 m. the mixed Jungle i.e.Pine, Oak, Semal etc. are found. Oak is valuable for fuel, fodder and agricultural use. The main neighbouring areas are Ramashram, Kothiyara, Bangar Dhar, Ghandiyal-dhar, Rududhar and lower limit of Saknidhar and Chiledidhar. Above 2500 m. the dense Deodar and other associated species are found such as Fir, Birch, Blue pine, Ringal (Arundinaria), Oak and Buras. These are the main heritage of the region.

On top sides where there is only a very small depth of soil, either of Erosion or the slope is too steep, tree growth is unable to maintain itself. In these places there is a growth of fine grasses. The temperate forest zone (above 1500 m.) provides favourable conditions for human habitations, consequently forest growth has been very greatly disturbed. The reserved forest, village community land is also under forests.

Types of Landuse

In the region, typical valley-section comprises the following categories of land which are on the basis of slope, soil, relief, irrigation means, pasture and vegetation.

(i) **Katil land:** On the higher parts of slopes, just below the forests, lie unterraced lands is called Katil or tilled land. China, Gahat, Bhatt, Jhangora and Madua are the most important crops in the region.

(ii) **Upraon land:** These are the permanent terraces but generally unirrigated cultivation is found in upland slopes. Important crops like wheat, dry paddy, madua, barley, Jhangora, China & oagal.

(iii) **Talaon land :** Land situated near the banks of rivers or rivulets (gadhera) is known as talaon land. Such type of land is known as 'Sera' and is invariable double-cropped with wheat and wet paddy.

TABLE 3: **Area Under Principle Crops (in Hects)**

Sub-division	*Rice*	*Maize*	*Madua*	*Wheat*	*Pulses*
1. Jakholi	3012	180	1496	4045	904
2. Kirtinagar	3920	1214	1550	5207	733
3. Deoprayag	4834	195	1920	5375	859
4. Narendra Nagar	3269	205	1250	4775	640

Source: Dist. Census Hand Book, 1981 (B).

Changes in Landuse

The land use changes in the region have been examined by the field studies. Most common land use changes are due to the heavy pressure on cultivated land. The inhabitants have constructed new fields in the cultivable waste land which are earlier utilised for fodder grasses and pasturing. The forests are being destroyed by the inhabitants. Therefore, the study area as well as the density of forest decreasing rapidly along the hill slopes. Cultivable waste land along the Mandakini, Bhagirathi, Bhilanganga, Alakananda and leester valleys (locally known as Baggar) are also being brought under cultivation. In rainy season, these Baggar are flooded. In many villages of the region there are changes in the Talaon land. Thus the area of talaon land is expanding regularly. Due to the construction of motorable road and canals, many problems have come into being. For example, roadside debris has silted the agricutural fields and the road construction has increased the number of landslides in the rainy season. The development processes in the region have created number of environmental problems.

REFERENCES

Singh, V.R., 1977. Taxonomy of Indian Agricultural Typology, U.B.B.P. Gorakhpur Vol. XII, 1977.

Maithani, D.D., 1983 Tehri-Garhwal- A Geographical Appraisal of Resources, Unpub. D.Phil. Thesis, Garhwal University, Srinagar.

Verma, R.K., 1985. Economic Development and Social Change in Doon Valley–Unpub. D.Phil, Thesis, Garhwal University, Srinagar.

Maithani, D.D., 1989. An Assessment of Nutritional Level in Tehri- Garhwal, G.G. J.I. Vol. III, 1989, Rishikesh.

— District Census Hand Book of Tehri- Garhwal, 1981 (B).